£2-25
F1

FLORA

of the Isles of Scilly

FLORA
of the Isles of Scilly

by

J. E. LOUSLEY

with drawings by

Barbara Everard

DAVID & CHARLES : NEWTON ABBOT

7153 5465 5

Set in Times
and printed in Great Britain by
Redwood Press Limited, Trowbridge & London
for David & Charles (Publishers) Limited
South Devon House   Newton Abbot   Devon

CONTENTS

LIST OF ILLUSTRATIONS

PLATES

TEXT FIGURES

LINE DRAWINGS BY BARBARA EVERARD

DISTRIBUTION MAPS

PREFACE

The Isles of Scilly are world-famous for their flowers, for the fields of winter-flowering Narcissi grown for the Christmas market, and for unfamiliar colourful and often succulent, exotic plants which give a subtropical aspect to the scenery. These are gardeners' importations growing there because the islands have the mildest winters in Britain. But the wild plants are just as remarkable. Some, such as the Dwarf Pansy and Early Adder's Tongue, do not grow on the mainland, and nowhere is there greater profusion of maritime plants. Scilly is a botanist's paradise.

Every year more and more people interested in flowers fall under the spell of the Isles of Scilly and so a *Flora* is long overdue. The purpose of this book is to record the wild flowers and ferns which have been found and their distribution in the various islands, and to offer explanations of why they are there.

In this it differs from the conventional local *Flora*. Islands are sharply defined areas with natural boundaries, and when they are small the interdependence of plants and animals on one another and on the activities of the human population is much clearer than on the mainland. History and land use have played as great a part in deciding the present plants in Scilly as geography, geology and climate. The patterns of variation and distribution of mammals, birds, butterflies and moths, beetles and grasshoppers etc. are all relevant. The study of Scilly has a wholeness which defies clear-cut division into disciplines and a mere catalogue of plant records would lose much of its interest.

There is so much interest in the Isles of Scilly that it is easy to forget the small size of the area. In most recent local floras the unit for recording is the 'tetrad', a square 2 km x 2 km as shown on Ordnance Survey maps. The *Flora of Hertfordshire*, for example, covers 335 complete tetrads and parts of a further 172; the whole of the land area of Scilly would be covered by less than four tetrads. Thanks to the irregularity of the coastlines and the small size of the fields, the work of investigating the flowers of Scilly is out of all proportion to the size, and the interest is endless.

1

It is from the small size that the greatest threats to conservation arise. Until recently the islands were protected from the influx of large numbers of visitors by the rigours of the sea crossing and lack of accommodation. Air services and the change from cut flowers to holiday facilities as the main industry have brought more people. In 1968 about 35,000 day-trippers and 40,000 longer-stay visitors came by boat, and a further 28,000 by helicopter, and the numbers are likely to increase.

Every day in summer many large boatloads of people are landed for the day on off-islands with potential damage from trampling, fires and other disturbance. In an attempt to meet the pressing need for more water, pumps have been erected in Higher and Lower Moors which are now much drier than formerly. Difficulties in disposing of so much rubbish have raised threats of using wet areas as refuse dumps.

The Duchy of Cornwall as landlords and the Isles of Scilly Council exercise strict control over development and, in the face of great pressures, they have been remarkably successful in maintaining the charm and interest of the islands. Commander T. M. Dorrien Smith has been compelled by economic circumstances to make changes on Tresco, but as lessee of the uninhabited islands he has continued the family interest in their wildlife. Landing on some, and especially Annet, is controlled and the flora is unspoilt. The Nature Conservancy shows increasing appreciation of the scientific importance of Scilly (I provided reports on the flora in 1946, 1954, 1957 and 1967) and the official outlook is encouraging. It is hoped that readers of this book will add their support. They are asked especially to refrain from picking the rarer plants and from trampling them when taking photographs. With such a small area to search they are easy to find.

The preparation of this *Flora* commenced immediately after my first visit in 1936. By 1940, with the willing help of a few friends, the distribution of most species was fairly fully recorded. During raids at the height of the London blitz a manuscript was typed ready for publication and sent to Cornwall for safety. Since the war this manuscript has been constantly revised to keep up with developments in botanical knowledge and name changes. With the help of the many botanists who now visit Scilly and my own field work, the six and a quarter square miles are now perhaps more thoroughly investigated than any other part of the British Isles of comparable size.

The collection of reliable records has involved extraordinary difficulties and it is as though a kind of midsummer madness is apt to descend on botanists when they land in Scilly. Decisions are easy when this takes extreme forms such as the recording of mountain plants, but much more difficult when the species claimed are common on the mainland but otherwise unknown in Scilly. Even recent publications include absurd records and some people for whose competence I have the greatest respect have proved unreliable in these islands. I cannot hope to have escaped errors myself but as far as possible this work has been based on herbarium material which can be checked if doubts arise. It has also been difficult to decide which records of alien plants justify inclusion. I have done my best to exclude hortal introductions deliberately planted on walls or elsewhere in the absence of evidence that they spread naturally. The scope for increasing the list with these garden plants is almost endless but would serve no botanical purpose. For many years I was Honorary Curator of the South London Botanical Institute in charge of the best collection of manuscripts and specimens of plants from Scilly and this, and the material at the British Museum (Natural History), Royal Botanic Gardens, Kew, and other leading herbaria, has been freely cited.

In writing this book I have had in mind people interested in wild flowers as well as dedicated botanists. A recent survey has confirmed statistically that a very high proportion of the visitors to Scilly have a keen interest in plants and wildlife and for their benefit I have given English names where appropriate.

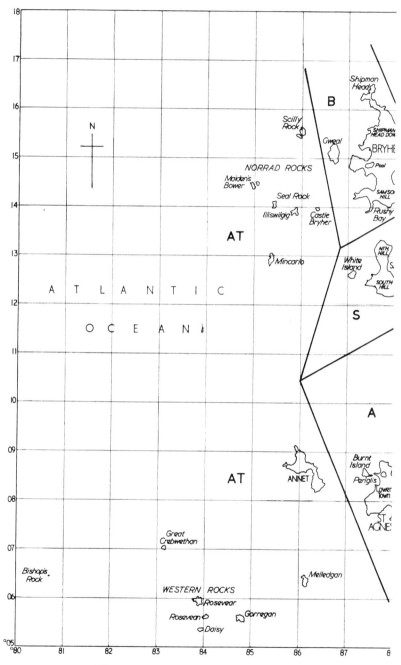

Fig 1 Isles of Scilly: Botanical Divisions

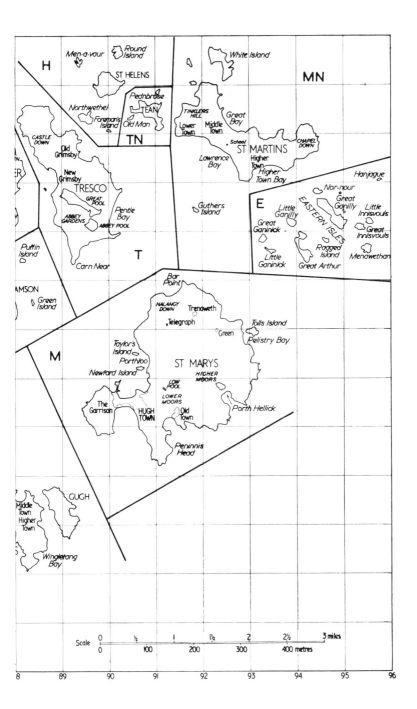

H

Men-a-vaur   Round Island

ST HELENS

Northwethel   Pednbrose
                TEAN
Foreman's   Old Man
Island

CASTLE
DOWN   Old
        Grimsby

New
Grimsby

TRESCO

GREAT
POOL

ABBEY   Pentle
GARDENS Bay
        ABBEY POOL

Puffin
Island

Carn Near   T

...AMSON

Green
Island

M

TN

White Island   MN

TINKLERS
HILL   Great
Lower   Bay
Town   Middle
        Town   School
                            CHAPEL
            ST MARTINS   DOWN

Lawrence   Higher
Bay   Town
            Higher
            Town Bay

Guthers   Hanjague
Island
                    E   Nor-nour
            Little   Great
            Ganilly   Ganilly   Little
                                    Innisvouls
            Great
            Ganinick   Great
                                        Innisvouls
                    Ragged
                    Island   Menawethan
            Little
            Ganinick   Great Arthur

Bar
Point

HALANGY   Trendweth
DOWN
    Telegraph   Green   Tolls Island

                            Pelistry Bay

Taylor's
Island
    Porthloo   ST MARYS
Newford Island   HIGHER
                        MOORS
            LOW
            POOL
            LOWER
            MOORS
The   Old   Porth Hellick
Garrison   HUGH   Town
            TOWN

                Peninnis
                Head

GUGH

Middle
Town

Higher
Town

Winkletang
Bay

Scale   0   ½   1   1½   2   2½   3 miles
        0   100   200   300   400 metres

8   89   90   91   92   93   94   95   96

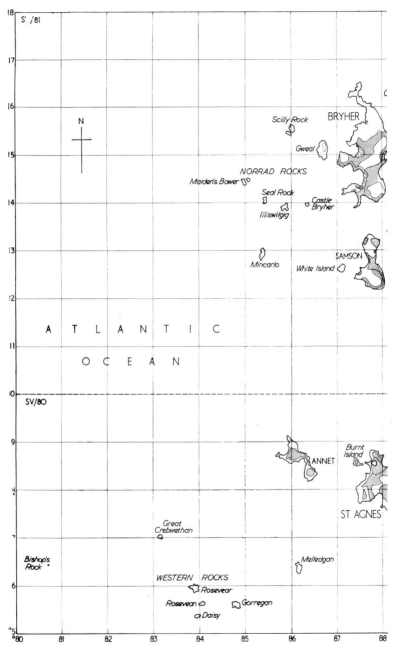

Fig 2 Isles of Scilly: Geology and National Grid

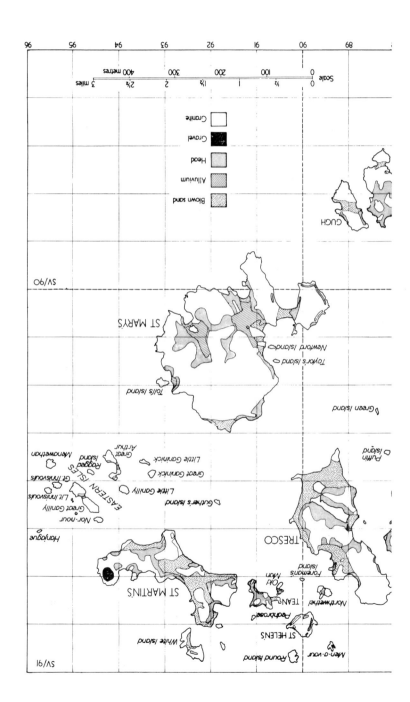

Chapter One

THE BASIS OF THE FLORA

The Isles of Scilly form an archipelago some twenty-seven miles. west-south-west of Land's End. At around latitude 49°56'N, longitude 6°18'W, they are the most westerly land in England and include the most southerly land in the British Isles. Their exposed situation in the Atlantic Ocean, and distance from the coast, ensure maritime features of great interest. Their only official name is the Isles of Scilly, except that 'otherwise Sully' is occasionally added to legal documents to cover an ancient spelling, 'Scilly' is thus acceptable as a shortened form, but 'The Scilly Isles' is to be avoided as objection-able and having official recognition only for a system of traffic islands inland at Esher, Surrey.

The islands are administered by the Council of the Isles of Scilly and constituted as a County Council area. Although there is some collaboration with Cornwall over such matters as police and educa-tion, they are completely independent for governmental and legal purposes. For botanical purposes they have been included in Watsonian Vice-county 1, West Cornwall, but for purposes of recording plant distribution botanists have found it an advantage to treat them separately as V.-c. 1 a.

There are five inhabited islands: St Mary's, Tresco, St Martin's, St Agnes and Bryher. The largest of these is St Mary's, which is about two and a quarter miles long. About forty smaller islands carry vegetation and are all now uninhabited. In addition there are about 150 named rocks which are too exposed or too small to form habitats for vascular plants. These islands and rocks form a gradual series from the largest inhabited island down to the smallest one with plants. This archipelago is arranged in an oval group some eleven miles in length from the Bishop Rock Lighthouse to Hanjague, and a little over five miles in breadth from Scilly Rock to Peninnis Head, St Mary's. It is set on a shallow shelf where the ten fathom line would include all but a few of the westerly rocks. Much of the sea between the islands is less than two fathoms (12 ft) deep.

In this *Flora* the islands have been used as units for recording

8

purposes and arranged in ten groups, as shown on the map (Fig 1) and Table 1. This is much more satisfactory than the conventional use of the National Grid, which would be misleading. The Isles of Scilly fall on four of the 10 km squares of the Grid—00/80, 00/81, 00/90 and 00/91, but there is so little on 00/90 that it is sometimes combined with 00/91 (Fig 2). The total land area is only 3963.2 acres (1603.56 hectares) which is less than one-sixth of the area of one 10 km square, so for the Distribution Maps Scheme of the Botanical Society of the British Isles all records from Scilly have been treated as belonging to one square (00/91) and the remaining squares ignored. Owing to the very irregular outlines of the islands the National Grid is hopelessly unsatisfactory for detailed recording. To avoid having more than one island on a square would involve using squares of 0·5 km or less and the practical difficulties would be great and the work excessive. By using islands as the basis for recording, and treating the sea as a natural boundary, the distribution of plants has been shown in a form which lends itself to comparison and interpretation.

Table 1 THE ISLANDS OF SCILLY

The areas shown are land above High Water Mark and were measured in acres from the 1963 editions of the 6in Ordnance Survey maps by the Nature Conservancy, and then converted into hectares. The headings are the botanical divisions used in this Flora and small offshore islands are associated with the larger ones to which they relate.

It has been necessary to calculate these figures on a uniform basis in view of the wide discrepancies between available statements of acreage. The reasons for these differences are complex but some of them arise because areas above HWM are taken for some purposes and LWM for others. There are differences between various surveys and there have been changes in parts of the coast.

INHABITED ISLANDS

	Acres	Hectares
St Mary's	1554	628.75
Taylor's Island	1	0.40
Toll's Island	6	2.43
Newford Island	1.2	0.48
	1562.2	632.06
St Agnes	270.1	109.28
Gugh	95.4	38.60
	365.5	147.88

Table 1—INHABITED ISLANDS—*continued*

	Acres		Hectares
TRESCO	735		297.38
		735.0	297.38
ST MARTIN'S	548		221.70
White Island	38		15.37
Pernagie Island	2		0.81
Plumb Island	1.3		0.52
		589.3	238.40
BRYHER	312.2		126.40
Gweal Island	15.2		6.15
		327.4	132.55

UNINHABITED ISLANDS

	Acres		Hectares
SAMSON			
Samson	95.4		38.60
Green Island	0.9		0.36
Puffin Island	2.0		0.81
White Island	3.8		1.54
		102.1	41.31
ANNET, WESTERN ROCKS & NORRAD ROCKS			
Annet	52.9		21.40
Melledgen	3.6		1.46
Gorregan	3.6		1.46
Rosevean	2.1		0.85
Rosevear	4.6		1.86
Great Crebawethan	2.1		0.85
Daisy	1.3		0.53
Scilly Rock	3.8		1.54
Illiswilgig	2.8		1.13
Castle Bryher	1.0		0.40
Maiden Bower	1.6		0.65
Little Maiden Bower	0.7		0.28
Mincarlo	4.5		1.82
		84.6	34.23
ST HELEN'S GROUP			
St Helen's	49.3		19.94
Round Island	9.5		3.84
Northwethel	12.0		4.86
Men-a-vaur	3.5		1.42
		74.3	30.06

Table 1—UNINHABITED ISLANDS—*continued*

	Acres		Hectares
TEAN			
Tean	39.5		15.98
Old Man	2.9		1.17
Pednbrose	1.4		0.57
		43.8		17.72
EASTERN ISLES			
Great Ganilly	32.6		13.19
Great, Middle & Little Arthur	19.0		7.68
Great Ganinick	4.5		1.82
Little Ganinick	2.2		0.89
Little Ganilly	6.0		2.43
Nor-nour	3.3		1.34
Great Innisvouls	3.8		1.54
Little Innisvouls	1.6		0.65
Menawethan	6.0		2.43
		79.0		31.97
TOTAL LAND AREA	*acres* 3963.2		*hectares* 1603.56

GEOLOGY

The larger islands of Scilly are a series of flat topped granite masses separated by shallow sea. Collectively they represent a dome-shaped granite mass (a granite laccolite) with the upper parts cut off by a plane of marine denudation. Granite is exposed, or covered with a very shallow soil, on all the higher ground, and it comprises all the sea-washed rocks. It forms the 'carns' or rocky hillocks, which are such a feature of the coastal scenery.

The granite of Scilly is of a rather special kind, whiter, more sparkling and softer than the rock of Land's End. It contains muscovite, biotite, orthoclase and quartz with other minerals. The charm of Scilly owes a great deal to this sparkling rock and the peculiarities of the flora are related to the soils to which it gives rise. There are two main types in the islands—one coarse-grained with porphyritic crystals of felspar, the other finer and non-porphyritic. The coarse-grained type is the more common and I am unable to detect any differences in the flora attributable to differences in the granite. In many places, and especially on Peninnis Head, St Mary's, and on the south-west side of St Agnes, the granite has been weathered into rocks of most grotesque shapes. The vertical and horizontal fissures are stages in the process of decomposition.

'Head', or 'Rab' is an accumulation of angular or subangular fragments of granite in an advanced stage of decomposition. It can be seen in many places on the cliffs and has been formed at the bottom of slopes from fragments of granite which have worked down as rock at higher levels broke up. Head forms quite extensive deposits in all the larger islands, and occurs also on some of the smaller ones. It provides a habitat for deeper rooted plants and especially Bracken, and the flora contrasts sharply with that of the shallow soils over granite on which Ling and Fine-leaved Heath are the characteristic species. On the inhabited islands most of the Head is under cultivation. Geologists distinguish Upper Head from the commoner Lower Head (both Pleistocene) but this appears to have no botanical significance.

Blown Sand represents a further stage in the disintegration of the granite and Head, and was formed by the pounding of sea waves on these rocks. As seen on the shore the sand of Scilly 'is essentially composed of finely comminuted granite, the dominant constituent now being minute grains of quartz, and white felspar; the latter imparting the snow-white colour' (Barrow, 1906). The vast amount of sand is out of all reasonable proportion to the present size of the land, and is explained by the denudation of the granite as the land level fell (see p 14). The sand formed from the submerged granite was thrown up on the shores and blown by the wind to higher levels, thus giving rise to the 'High Level Blown Sand'. Along the south and much of the eastern shore of Tresco, along a considerable part of the north and south coasts of St Martin's, across Old Town Bay, St Mary's, and elsewhere, the sand has been raised up into practically continuous ridges just above high-water mark which serve to keep out the sea. In places the sand has blown up to cover quite high ground—it goes right over the ridge of St Martin's—and it goes from shore to shore to join up parts of islands as at Hugh Town, St Mary's, and Samson, and the Gugh to St Agnes. Until Augustus Smith, the Lord Proprietor, took steps to arrest it by planting Marram Grass and Hottentot's Figs soon after his arrival in 1834, white blown sand drifted in high winds to cover crops and penetrate houses.

Blown Sand is the basis for the soil in several areas of great botanical interest—Bar Point and Pelistry Bay on St Mary's, Appletree Banks on Tresco, the neck of Samson, Rushy Bay, Bryher, and

Great Bay on St Martin's, are a few examples. It also affords excellent ground for bulb cultivation, and especially when on a south-facing slope.

The island of St Martin's may be taken as an example of the close relation between geology and cultivation. Examination of the 6in geological map shows that the tiny enclosures which are the arable fields coincide almost exactly with the areas of Head or Blown Sand. The larger fields, which are rough grazing, are on granite. The sandy slopes above the two large bays on the south side of the island are veritable sun traps, and the warm sandy soil produces the earliest flowers in all Scilly.

Two patches of alluvium were recognised by the Geological Survey: at Higher Moors, and Lower Moors, St Mary's. The latter (Old Town Marshes) is protected from inundation at spring tides by bars of Blown Sand at Porthloo, Porth Mellon and High Town Bay. Before these barriers were built up, sea flowed through the gaps and filled the greater part of the area within with storm-beach boulders. Over this a fine clay not unlike china clay in tenacity, and usually less than two feet in thickness, has formed from material brought down from the neighbouring high ground. For a time the site of Lower Moors was probably a lagoon in which aquatic vegetation assisted in building up the alluvium as the water became progressively less brackish. It is likely that *Phragmites* swamp followed, and this still persists in small areas. The embankment and road at Old Town, and the construction of an artificial drain, have hastened the process of excluding the sea and making the marshes drier but the whole development is relatively recent. Lower Moors (and Higher Moors) are marked as 'maresh' with fresh water on a map made about 1585 (Cottonian MSS Aug 1, vol ii. 18), but the sea is still liable to break in when spring tides coincide with high winds in the appropriate direction.

The other area marked as Alluvium, Higher Moors, has a similar history but has reached a less advanced stage. The bar at Porthellick keeps out the sea, but the Fish Pond has nearly five acres of open water, fringed with *Phragmites*. It is fed to the north by a small stream which still brings down alluvium, though its rate of flow is very slow.

There are three more geological rocks that deserve brief mention. Associated with the Head is 'iron cement', a sandy deposit, locally called 'clay', which dries so hard that it has been used in old buildings as mortar. When set a thin layer will bear the weight of a man and it is used for the construction of roads. The botanical interest of this deposit is that it appears to form the water-proof beds of the fresh-water lakes on Tresco, and perhaps also of the freshwater pools on Bryher. It contains soluble silica and a large amount of iron oxide and these explain its cement-like qualities. Barrow points out that in the interior of the islands 'a thin cover of it occurs in many of the hollows intervening between the more or less bare patches of granite, but it is too thin and irregularly distributed to be mapped out'. The effect seems to be very similar to that of an iron pan. A high-level gravel deposit on St Martin's at about 150 feet was regarded by Barrow as 'perhaps of Eocene date' but G. F. Mitchell (1960) gives convincing reasons for treating it as a Lower Pleistocene aggradation deposit. On White Island, St Martin's there is a small deposit of tourmalised schist (killas), but this, like the gravel, has no botanical significance, although it is of great interest to geologists.

CHANGES IN SEA LEVEL

The Isles of Scilly have been separated from Cornwall by sea for at least 300,000 years. In the Lower Pleistocene they may have been more extensive than they are today and connected to the mainland, but by the Middle Pleistocene they were detached and surrounded by sea. The glaciations of the Middle and Upper Pleistocene did not reach them (Mitchell, 1960).

The changes which have taken place within the archipelago have considerable botanical significance. Crawford, in his classic paper of 1927, suggests that it may be taken as proven that the Scillies were once a single large island—or one big one with a few islets on reefs. Much additional evidence has come to light since his paper was written and it is now possible to be a little more confident about the timing of the submergence.

At the present time the seas between Bryher, Tresco, Samson, and St Martin's are so shallow that people frequently walk between them at spring tides. On a single tide on 25 September 1961 Mr J. Pickwell starting from Samson crossed on foot to Tresco and then on to St Martin's—a total distance of 4.9 miles. A rise of only 30 ft would unite these four islands with St Mary's, Tean, St Helen's and the Eastern Isles. A rise of 60 ft would join St Agnes and the Western Rocks to the rest.

Submerged stone walls (hedges) have long been known on Samson

Flats (Borlase, 1754; Crawford, 1927). They are also reported from New Grimsby Channel (*Scillonian* **21**, 74) and Tean Flats, where their nature as field walls has been confirmed by the discovery of rotten turf beneath them. A village of seven huts was found by undersea divers in following the submerged causeway from near Bar Point, St Mary's towards St Martin's, and it has been suggested by Leachman that this causeway is likely to be later than 1066 (*Scillonian* **32**, 114; **37**, 43–45; **20**, 44–48).

Many prehistoric sites have now been found at or below sea level. For example, remains of Bronze Age habitations at English Island Carn, St Martin's; Halangy Down, St Mary's, and on Nor-Nour may be broadly dated at 2000–1500 BC, and there was evidence of a mixed farming economy (Ashbee, *Scillonian* **170**, 149–157). A house partly uncovered by the sea at Little Bay, St Martin's was in use at the very end of the Bronze Age, *c* 500 BC (*The Times*, 27 August, 1953). At Par Beach, St Martin's there were houses of the Middle Roman Period, AD 200–300, covered by the sea at high tide, and further reports of Nor-Nour finds ranged from 1000 BC to AD 371 (*Scillonian* **164**, 196). When these habitations were constructed the land must have been very considerably higher in relation to the sea than it is at present.

The evidence suggests that in the Middle and Late Bronze Age there was a considerable area of land within the ring of islands which was mainly Head, fairly level, easy to cultivate and fertile (clay has been found below the sand). Corn was grown and the land cultivated by the people who buried, and later cremated, their dead in the tombs still to be seen on the higher ground of the present islands. Part of the area was probably marshland and possibly some oak-hazel wood.

Part of the area is likely to have been submerged at the same time as the Mounts Bay oak-hazel wood went under the sea and this de Beer dates with confidence at about 1700 BC as a result of the melting of the antarctic ice cap (*Scillonian* **167**, 100). It may be that this submergence cut off St Agnes but still left the other islands joined, since the archaeological evidence cited indicates that the flats off Tean, St Martin's and the Eastern Isles were dry land much later than this. Submergence may have been gradual and continued well into Celtic times before coastlines approximated to the present ones. Even in very recent times further land has been lost. It is alleged that within living memory men worked in the fields of what is now Pilchard Pool, Porthcressa (*Scillonian* **20**, 65), and agricultural land in the same area was lost early in the nineteenth century (Woodley, 1822 and Forfar, 1896). There has been recent encroachment in Bryher, and the size of the island of Arthur is reduced (Mumford, 1967), while Brewer in 1890 attributed a reduction in agricultural acreage to the inroads of the sea.

The botanical implications of this are important. Separation from the mainland and freedom from glaciation dates back so long that considerable differences may be expected between the flora of Scilly and that of the adjacent parts of Cornwall (see p 43). St Agnes has probably been separated from the other islands before they ceased to be joined, and this could be significant. The rest of the islands have been joined until quite recent times and differences in their floras due to historic factors are not to be expected. The presence of land more suitable for cultivation now submerged enabled Bronze Age peoples to exist in greater numbers. They probably used much of the present islands, and almost all the rocky islands now uninhabited, for pasture rather than cultivation. The soil and vegetation were therefore less disturbed than would otherwise be expected.

CLIMATE

The special features of the flora of Scilly are closely related to climatic factors. The 'sub-tropical' gardens and the winter bulb industry are obvious examples. The native flora similarly includes species which will not thrive on the mainland, and others which flower 'out of season'. While statistics based on averages, such as those which appear in geography textbooks, give a useful overall picture of the climate, the botanist is more concerned with maximum and minimum figures collected over a long period. The persistence of species, and especially of those which are less adaptable, is determined by climatic extremes which may occur only at long intervals.

The climate of the British Isles as a whole is usually described as maritime or oceanic and the Isles of Scilly are of the extreme Atlantic type, shared only with places exposed to the full influence of the Atlantic, and even milder than the climates of the west and south coasts of England. Thus the climate of Scilly is characterised by a most equable temperature, adequate and regular rainfall, near freedom from snow, frequency of sea fogs, frequency of strong

Table 2 THE CLIMATE OF THE ISLES OF SCILLY

(Station: St Mary's, 158 ft above mean sea level, with averages for periods stated.
Compiled from figures supplied by the Meteorological Office)

	Jan	Feb	Mar	Apr	May	Jun	Jul	Aug	Sep	Oct	Nov	Dec	Year
TEMPERATURE °C (1931–60)													
Daily Maximum	9.2	9.0	10.5	12.1	14.3	17.0	18.7	19.2	17.5	14.6	11.8	10.0	13.7
Daily Minimum	6.3	5.7	6.5	7.4	9.2	11.7	13.4	13.7	12.9	10.8	8.7	7.2	9.4
Mean	7.7	7.3	8.5	9.7	11.7	14.3	16.0	16.5	15.2	12.7	10.3	8.6	11.5
Extreme maximum	13.9	13.3	16.1	19.4	21.7	26.1	25.6	27.8	24.4	20.6	16.1	13.9	27.8
Extreme minimum	−3.9	−3.3	−1.7	1.7	2.8	6.1	9.4	8.9	6.7	3.9	2.2	−2.2	−3.9
SUNSHINE in hours (1931–60)													
Total	62	81	130	192	235	228	207	208	155	121	76	57	1752
Daily mean	1.99	2.87	4.19	6.41	7.56	7.59	6.68	6.72	5.16	3.90	2.54	1.84	4.80
RAINFALL in inches (1916–50)													
Total	3.55	2.72	2.52	2.11	2.17	1.74	2.22	2.52	2.46	3.57	3.61	3.52	32.71
RAIN DAYS (more than 0.01″)	21.9	17.1	16.1	12.9	13.7	13.3	16.1	15.6	15.8	17.1	19.5	21.6	200.7
AVERAGE NUMBER OF DAYS (1931–60) OF													
Snow or sleet	1.5	1.2	0.6	<0.1	<0.1	—	—	—	—	—	<0.1	0.5	3.9
Snow lying (at 0.9 GMT)	0.3	0.3	<0.1	—	—	—	—	—	—	—	—	0.1	0.8
Air frost	0.9	1.9	0.8	—	—	—	—	—	—	—	—	0.3	3.9
Gale (Force 8 or over)	3.6	2.2	1.3	0.7	0.4	0.3	0.2	0.5	0.7	1.1	1.7	3.6	16.4
Fog	0.8	0.7	1.7	1.0	1.4	1.9	2.3	1.3	1.2	1.3	0.7	0.4	14.6

17

winds, and an excellent sunshine record (Table 2). The influence of all these is reflected in the flora, and they must be considered in detail.

TEMPERATURE

The temperature is remarkably equable throughout the year, with only a 9.2°C difference between the average of the hottest month (August 16.5°C) and the coldest (February 7.3°C). The months of December, January and February in Scilly are almost as warm as a London April, so that in normal winters plant growth in places sheltered from the wind is hardly interrupted. Scilly shares with the extreme west of Ireland an average of over 350 days in the year with an air temperature above 5°C and this is a rough measure of the days during which plants can grow.

The normally high winter temperatures have a great influence on the flora. This is clearly shown in Tresco Abbey Gardens where many handsome species from much warmer lands than our own, which will not grow in the open throughout the year anywhere else in England, not only thrive with only occasional protection, but in many cases produce their flowers in winter. It is also demonstrated by the flower industry which produces blooms from November to March, often weeks ahead of the mainland growers, and including southern narcissi, such as the Soleil d'Or which cannot be grown elsewhere (see p 36). Similarly most of the weeds of the bulbfields are species which make their growth during the winter and flower in the spring—some of them aliens from the Mediterranean. Many native plants continue to grow and flower in the winter months and in early spring there is a strange mixture of blooms associated on the mainland with other seasons. For example, in March I have been surprised to see Fine-leaved Heath, *Erica cinerea* and Long-rooted Catsear, *Hypochoeris radicata* in full flower. Some flowers appear exceptionally early—such as Coltsfoot, *Tussilago farfara*, which has been seen in several years in the first half of December.

The flora includes an exceptionally high proportion of species which are active in winter. Some, like *Allium babingtonii*, which comes up in January and dies down in June, and *Arum italicum*, which appears above ground in winter and dies down in July, are adapted to avoiding the summer months with their droughts as the inclement period of the year. Some species with a southern distribution grow in Scilly at the extreme of their range and are sensitive to the rare frosts. *Lavatera cretica*, for example, often germinates in autumn and, if the winter is a normal one, flowers in May. Following exceptional cold winters visitors in early summer fail to find it. *Ornithopus pinnatus* normally grows through the winter, but is killed by frost in exposed places. In such years, which only occur rarely, the stock is replaced from seed.

It is the rarity of severe winters which enables marginal species to survive, and a sequence of years in which frost occurs might well cause some of them to become extinct. The lowest temperature recorded in Scilly is −4°C, and cold conditions generally arise from a north-east wind, from the direction of the nearest land. This condition prevailed, for example on 29 January, 1947 when the maximum temperature was −2°C, although the islands were surrounded by sea with a surface temperature of nearly 10°C. On the same day the maximum temperature at Plymouth was only −5°C. Drifts of snow formed up to 4 ft deep and stayed on the ground for five days. The Pelargoniums and succulents which are grown so freely were killed in many gardens. Other severe winters have been 1891, when there was enough snow to make snowmen; 1940, when the minimum temperature on 20 January was only −3°C; 1954, when the temperature dropped to −3°C on 2 February, and pools froze; 1956, when 5° of frost were recorded; 1957, when there was a hoar frost on 1 January; 1962, when there was snow on 27 February; and the winter of 1962–3 which is remembered for the very long cold spell which lasted from 22 December to 5 February, with many days of frost, and snow on the ground for three days.

These are the worst conditions that the plants of Scilly have had to face over the last eighty years and the damage caused has been remarkably little. I was in the islands in the summer months following the severe winters of 1940, 1954, 1956, 1957, and 1963 and it was very evident that the vegetation harmed was mainly that exposed to north-east winds. The best widespread indicator of frost damage is the South African Hottentot's Fig, *Carpobrotus edulis*, which is cut back and even killed by frost. Even in the worst years much of it escaped damage when sheltered by higher ground or growing in the crevices of rocks. The distribution of this species within the islands suggests that it needs shelter from winds from the north-east. My observations indicate that the damage depends on the precise direction of the wind as much as on the number of degrees of frost during the cold spell, and this probably explains why the native plants showed little evidence in May 1940 of the exceptionally cold conditions earlier. Even under the worst conditions yet experienced it is likely that considerable areas are sufficiently sheltered for even the most tender native species to avoid serious setbacks.

High temperatures are as uncommon as low. The extreme maximum recorded is 28°C, but figures approaching this are exceedingly rare. When the temperature reached 22°C in July 1961 the *Scillonian* magazine said the heat 'was almost unbearable'. Plants which require sustained high temperatures to mature seed are unlikely to reproduce in Scilly.

In Scilly the drop in temperature at night is very much less than on the mainland. In winter and spring especially, this extends the number of hours when plants can maintain active growth and no doubt assists species with a very short life cycle, such as *Poa infirma*, and the rapid development of flowers in the bulb industry.

RAINFALL AND HUMIDITY

The average rainfall of Scilly is 32.71 in, which is very low for the south-west, but near the average of 32.67 in for the whole of England. Rainfall is heaviest from October to January, and at its lowest in April, May and June. Even then the monthly average does not fall below 1.7 in. The numbers of days on which measurable rain falls averages 200, and in no month is there an average of less than 13 rain days. Very little precipitation falls in the form of snow, the average number of days on which snow falls being less than four per annum—about the lowest in the British Isles. When it is realised that only a few minutes of snow or sleet falling between midnight and midnight count as a 'day with snow', and that a few minutes about midnight will count as two days, it will be appreciated how extraordinarily low this figure is.

The even spread of the precipitation and the high humidity (see below) ensures that under normal conditions plants are seldom short of water. It also explains a feature which puzzles many visitors, the occurrence of plants in much drier habitats than is usual on the mainland. As Townsend remarked of Bog Pimpernel, *Anagallis tenella*, 'Very common even in apparently the most unlikely situations, showing the dampness of the climate; otherwise, the soil being so shallow, it would soon become parched' (Townsend, 1864 p 15). Other species of which this is true include *Radiola linoides* and *Parentucellia viscosa*. On the rare occasions when there are long periods without rain, such as in the glorious summer of 1955 when drought lasted for eight weeks, plants on the shallow porous soils of Scilly suffer very severely. Large areas are also devastated by heath fires (see p 41). It is very rarely that the even spread of the precipitation through the year is sufficiently upset to cause flooding, but this happened in 1947. The melting of the January snow turned Lower Moors, St Mary's into a big pond, and as late as May the Moors were still a sheet of water with reeds standing up here and there.

Humidity averages for 1921–1935 are set out in Table 3.These show that the moisture content and relative humidity are high throughout the year. The relative humidity varies only between 81 (June, July)

Table 3   AVERAGES OF HUMIDITY

From *Averages of humidity for the British Isles*
Met.O.421.   HMSO   1938

	Scilly Cornwall 49°56N., 6°18′W., 163 ft. 1921–1935				Relative Humidity	
	Averages for 13h				Relative Humidity	
	Temp	Rel Hum	Vap Pressure	Moist Content	7 h	18 h
	°F	%	mb	gm/m³	%	%
January	48.3	86	9.8	7.7	89	88
February	47.3	85	9.4	7.3	89	87
March	49.1	84	9.9	7.6	90	87
April	51.1	82	10.5	8.0	89	85
May	55.1	83	12.2	9.3	90	86
June	60.4	81	14.5	10.9	91	85
July	63.8	81	16.4	12.3	91	85
August	63.6	82	16.6	12.3	93	86
September	61.4	83	15.3	11.4	92	88
October	56.5	83	12.9	9.8	89	88
November	51.1	83	10.5	8.1	87	86
December	48.7	84	9.9	7.6	88	88
Year	54.7	83	12.3	9.2	90	87

and 86 (January) on the averages for 1 pm, and the variation between the morning and evening figures is very small. The frequency of fogs—14 days in the year—is a reflection of the high humidity of the air, and the influence of the sea, and the fogs are most common in the summer when the moisture content is greatest.

LIGHT

The sunshine record of Scilly is good (see Table 2). With an average over the year of 38 per cent of the possible maximum, rising to 46 per cent in May and June, with an average of nearly 8 hours a day, these are near to the highest figures in the British Isles. It is likely that the effect on plant life of these long hours of sunlight is enhanced by the exceptional quality of the light. Owing to its situation the atmosphere over Scilly is almost free from dust and industrial pollution, and this, with reflection from the shallow sea and sandy beaches, results in exeptionally high proportions of ultra-violet rays. These, which cause so much pain to visitors liable to sunburn, and so many over-exposed photographs, probably have a considerable influence on the flora. The strong light is likely to favour species with a Mediterranean type of distribution and it is thought to be responsible for the enhanced brightness in the colour of the flowers of some species.

WIND

The favourable factors to plant life are to some extent offset by the frequency of strong winds. The wind velocities recorded in Scilly are amongst the highest for the British Isles, including one of 111 miles per hour recorded on 6 December, 1929. During the five years from 1911 to 1915 there were no fewer than 637 observations of winds with a velocity of more than 24 miles per hour, and most of these blew from the west. The frequency of strong winds is well shown from figures given by Robertson. These, prepared from averages taken over ten years, show the number of hours per annum for winds at the velocities stated and the following is an extract (Robertson, 1941):

Over 39 mph	25–38 mph	13–24 mph	4–12 mph	0–3 mph	No record
160	1718	4140	2400	329	3

A few examples from recent years indicate the forces which winds attain in Scilly and their effect. In 1965 there were gales of 83–85 mph in January which whipped the seas up to 100 ft, and gales of about 80 mph in September. In 1964 there were gale force winds in September followed by a 70 mph gale on 27 December. On 15 March 1963 there was a hurricane of 95 mph. In 1962 there were gusts at the Telegraph up to 90 mph in January and of 85 mph on 7 March. This gale broke sea walls, washed away parts of the cliff and washed bulbs out of the ground. There were further gales that year of 60 mph on 18–19 May, of 70 mph on 29 September, and one with gusts of 70–80 mph at the end of December. In 1961 there were five gales of over 50 mph spread through the year. In 1954 there were continuous gales from 29 November to 16 December. Wind velocities of 97–100 mph were recorded at the Telegraph, the Round Island instrument smashed at over 110 mph, and the Bishop Rock reported 110 mph with seas racing past the window 90 ft up in the tower. Pine trees were uprooted and crops destroyed.

Gales are most frequent in December, January and February but, as these examples show, they are not restricted to the winter months but occur throughout the year. They are often accompanied by tremendous seas, with waves sometimes 90 or 100 ft high. The salt-laden spray from these waves is driven by the wind right over the islands so that, while the influence of salt is greatest on the western side from which the gales come, there is hardly a spot in the archipelago free from the effect of salt spray. The smaller islands are completely under its influence (see p 53). Species associated with saline conditions, like Sea Spleenwort, *Asplenium marinum* (plate p 193), and Danish Scurvy-grass, *Cochlearia danica*, thrive away from the sea even on high ground. On the other hand, plants which cannot tolerate salt are unable to grow in Scilly, and the shrubs to protect the bulbfields are all chosen for their ability to stand these conditions.

In addition to carrying salt, winds have a limiting effect on plant growth since they increase water loss from soil and vegetation, have a cooling effect, and cause mechanical damage. In Cornwall trees are killed by the loss of water in windy places. In Scilly they cannot grow at all unless they have protection. In an interesting old manuscript dating from about 1695, the writer describes how trees were successfully planted at Holy Vale in the centre of St Mary's, which is the most sheltered spot in Scilly. He goes on:

'Eight Appletrees, planted by Mr Tho. Child near 30 years agoe at Tremalathan in the same Island; the thickest whereof are not yet a foot and ¼ in circumference nor the highest above 10 foot high. They throve very well, kindly and fatt, till they were growne to the height of the Wall wherewith they were fenced about, but there they stopt and have been at a stand for severall years, they being now but little higher than the top of the said Wall. This he attributes to the Northwinds, which being there extreame cold frequent and high check and hinder their growth; adding that noe trees will grow here unless fenced and sheltered from those Winds. . . . ' (Turner, 1964).

Except that winds other than from the north are commonly concerned, this is a graphic description of the experience to which many writers have drawn attention. Trees will only grow in Scilly where sheltered from the wind and even then it has been necessary to use gorse and other bushes as 'nurses' to give them a start. Similarly with crops, 'hedges' and 'fences' must be provided to protect them from the wind (see p 37).

The influence of wind on the native vegetation is very great indeed. The stunted *Callunetum* on the higher ground of Bryher, Tresco and St Helen's and on lower ground on the west side of St Agnes is an excellent example of this. Here Ling, *Calluna vulgaris*, can be seen as gnarled woody plants rising only a few inches above ground level. The windward side of the bushes shows only dead wood, while the living shoots grow out in its shelter on the leeward side. In such windy places every granite boulder provides protection for plants, and it is also common to find gorse (*Ulex* sp) bushes grown out in the form of a cushion in the shelter of small boulders (Fig 3).

The work of the wind in moving sand and piling it up on higher ground has already been mentioned. Although most of the dunes

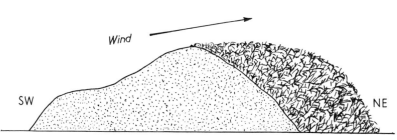

Fig 3 Wind Pruning: Gorse, *Ulex europaeus*, in the shelter of a granite rock on the Garrison, St Mary's. (After Leuze, 1966.)

have now been fixed by planting suitable sand-binding species, there is still considerable local disturbance after gales when plants have to grow up through the sand. Because of the wind, shelter is all-important to plants in Scilly and produces marked differences in microclimates. Thus the south end of Tresco is usually a few degrees warmer than St Mary's, and Tresco Abbey Gardens are often very much warmer than just outside.

COMPARISON WITH CORNWALL AND THE CHANNEL ISLES

It will be shown later that the flora of Scilly differs greatly from that of the nearest mainland, whereas there are close resemblances to the flora of the Channel Isles. The figures for temperature and rainfall on St Mary's have been compared by Robertson (1941) with those for Gulval, a sheltered place on the outskirts of Penzance. It is only during the winter months, from November to March, that the average temperatures show a significant advantage in favour of Scilly but in every month of the year the minimum temperatures are higher. For rainfall the difference is surprisingly great. In every month Gulval has more rain, with an average total for the year of 40.9 in compared with 31.7 for Scilly.

More recently, the climate of Scilly has been compared with the adjacent mainland on thirty years averages obtained by W. L. Rees (*Scillonian* 27, December 1952) who gave the following figures:

	Sunshine (hours)	Rainfall (inches)
Scilly	1708	31.96
Penzance	1693	40.93
Plymouth	1677	36.72

He also found that throughout the winter months Scilly has the highest average temperatures, but in summer the islands are two or three degrees Fahrenheit below the mainland. The effect is to give winter growing plants an advantage in Scilly where they have to contend with less severe falls in temperature.

The Channel Isles also have mild winters, though not quite as mild as Scilly. Air frost is recorded at St Mary's on an average of four days a year, in Jersey on eleven. Snow lies twice as often in Jersey, where extreme minimum temperates of $-8.3°C$ are recorded for both January and February compared with $-3.9°$ and $-3.3°C$ in Scilly for those months. These differences control which of the more tender species are able to grow and persist, and explain why succulents so abundant in Scilly are few in the Channel Isles. Hottentot's Fig, which suffers in Scilly only in exceptionally severe winters, is commonly damaged by frost in Jersey. In Scilly the season for harvesting narcissi in the open is 10–14 days earlier than it is for the same varieties in Jersey. In view of the exchange of stock bulbs, these figures have almost the precision of a scientific experiment.

For high summer temperatures and sunshine the advantage is with Jersey. Here the summer averages are higher, and the extreme maximum (in August) is as high as 35.6°C. St Mary's in 1931–60 did not achieve a temperature higher than 27.8°C. While little detailed information is available on the effect of high temperatures on native species, it is evident that in most years annuals and other shallow rooted plants are liable to be burnt up more often and more severely in the Channel Isles. Some species which grow there are probably absent from Scilly because summer temperatures are not high enough to allow them to mature their fruits.

For plant life the Isles of Scilly offer conditions unparalleled in Britain. An archipelago built up of granite rocks subject to the full influence of the Atlantic with its gales and mild winters. This is reflected in the native flora just as it is in the bulb industry and the horticultural wonders of Tresco Abbey.

Chapter Two

MAN'S INFLUENCE

There is no other part of England where the influence of Man on the flora can be studied better than in Scilly. The islands present a series ranging from the larger ones where human influence is at a high level, down to smaller uninhabited islands less disturbed than anywhere in mainland England. Yet even the smallest islands have not always been left to nature, and these will be discussed later. It is the purpose here to outline sufficient of the history of the archipelago to show how trade and land use have influenced the flora generally.

PREHISTORY

The earliest inhabitants of whom we have definite information were the Bronze Age people who built the megalithic tombs still to be seen. These people colonised Scilly between 2000 BC and 1500 BC. They lived 'in small villages, cultivating wheat and barley in their small irregularly shaped fields tending herds of oxen and flocks of sheep (with some pigs and goats) about their farmsteads. . . .' (Daniel in *Scillonian* **21**, 116–118, 1947). Over fifty of their megaliths occur in Scilly which, with four in West Cornwall and five in Co Waterford, are distinct from other burial chambers in the British Isles. These people probably colonised Scilly from West Brittany after working their way up the west coast of Europe from the Mediterranean. Professor Daniel suggests that some at least of the old field walls on the islands, and submerged on the flats between them, are as old as the chambered tombs building period and 'may well mark out the cornplots of the megalith builders'. At this period large areas between the islands were dry land (see p 15) and the soil, having been formed from downwash from higher ground, was probably flatter and deeper than it is over most of the islands now. It seems probable that Bronze Age man used part of this for his crops and the higher rocky ground as pasture.

Thus corn has been cultivated in Scilly for nearly 4000 years, and initially by people who had originated from the Mediterranean and

27

could have brought Mediterranean weeds as impurities in their seed corn. It seems also that they continued to trade with the Mediterranean, as faience beads dated from the nineteenth Egyptian Dynasty (*c* 1320–1200 BC) have been found in one of their tombs (O'Neil, 1952). Charred wood and charcoal of oak and other trees has also been found, with evidence that they dried their grain before storing it in large jars. They must also have used a lot of fuel for cremating their dead. It seems that there was woodland, which may have been on part of the land now submerged, and Paul Ashbee suggests that the first crops may have been raised on land cleared by fire from forests (*Scillonian* **170**, 149–57, 1967).

In the last century BC there was a period of Breton-Cornish commerce in which the Veneti, a maritime tribe of Southern Brittany, were active. They are thought to have been responsible for building the Giant's Castle and other earthworks in Scilly and Cornwall (Wheeler, 1941) and their trade would have provided opportunities for the introduction of plants from north-west France.

In Roman times the mixed farming economy continued, with improved mills for grinding flour, and higher cultural levels. A Romano-British hut found at Par Beach, St Martin's in 1948 yielded 'Roman pottery of the third and fourth centuries AD and pieces of casserite or tin ore. The floor of this hut is now just covered by normal high tides. If one allows that in the fourth century it lay twenty feet or more above high-water mark, one is entitled to consider that the present St Mary's, St Martin's, Tean, St Helen's, Tresco and Bryher at that time still formed part of one island.' (O'Neil, 1949). This is confirmed by the position of other finds. The ancient village at Halangy, dated at about AD 200, and the house with Roman remains on Nor-Nour both produced evidence of cereal cultivation (Ashbee, 1955 and *Scillonian* **37**, 145, 153, 1962). Whether they came for tin, or just used the islands as a staging post, the Romans were evidently active in Scilly.

EARLY HISTORY

There follows a long period with only scrappy information about the history and almost nothing to throw light on the land use. There was a fifth or sixth century Christian settlement on the site which became Tresco 'Abbey', and a celtic church on St Helen's. Vikings used the islands as a base. Then a monastery was established a little before the Norman Conquest, and in 1114 Henry I made a grant of

Tresco and the neighbouring islands to the Abbot of Tavistock, and a Benedictine priory dedicated to St Nicholas was established on Tresco. Scilly gradually sank into a state of lawlessness and became the resort of pirates. By the mid-fifteenth century the 'Abbey' was in a state of disrepair, and long before they finally left at the Dissolution in 1539, the monks had ceased to have any importance in Scilly. They probably left their mark on the flora by the introduction of Soapwort, *Saponaria officinalis*, and Tansy, *Chrysanthemum vulgare*, which can still be seen on Tresco. Elecampane, *Inula helenium*, which was formerly known near Old Town, was probably introduced for medicinal purposes at about the same period. Since most of the narcissi which formed the basis of the early bulb industry were found wild round the Abbey (Gibson, 1934) it has been suggested that the monks were also responsible for their introduction.

In 1200 King John granted to the monks the 'tythe and three acres of assart land in the Forest of Guffer'. Assart was arable land made by clearing tree stumps and scrub from forest and the site of the forest of Guffer or Guffaer is not known. Even as late as this, it is possible that the forest was on land between the islands now under the sea, or it may have been in Tresco Wood (above the Abbey) where there were large tree stumps before Augustus Smith planted it up as shelter for his gardens.

For a time land was going out of cultivation. John Leland, librarian to King Henry VIII, in his account of the islands *c* 1535 describes the exceptional fertility of the soil of St Mary's for corn and goes on to say 'Few men be glad to inhabite these islettes, for al the plenty, for robbers by the sea that take their catail by force. The robbers be Frenchmen and Spaniardes'. (Leland, edit Smith, 1907). In 1579 Francis Godolphin reported 'There are now not a hundred men but more women and children; the tillable ground does not find half of them bread. Only the two islands wherein are fortifications [St Mary's and Tresco] are inhabited . . .' (*Calendar State Papers, Elizabeth, Add.* 1566–1579). The recognition of the strategic importance of the islands led to the building of the Star Castle and other fortifications. In spite of wars and other vicissitudes the population (excluding military) increased to 1122 in 1715, about 1400 in 1750, and the highest ever of 2618 in 1938. By this time every scrap of land capable of cultivation with the tools and under the conditions of the time, had become arable.

The general picture is clear—much land broken up in the Bronze

Age, Romano-British, and perhaps Celtic periods went out of cultivation and many of the weeds and other plants which had been introduced were probably lost. From Elizabethan times the trend was to take more land into cultivation. By 1838 there was a lot of arable which is no longer cultivated today.

FROM 1831

In 1831 there was a change in Lord Proprietors which had a major influence on the economic history of the islands and their flora. The Godolphin family surrendered the lease they had held since 1571, and in 1834 the Duchy of Cornwall granted a new lease to Augustus Smith for a term of 99 years. It is to the energy and initiative of this wealthy Hertfordshire squire and his love of horticulture shared by his successors T. A. Dorrien-Smith and Major A. A. Dorrien-Smith that we owe much of the present face of Scilly.

For purposes connected with the change in proprietorship, accurate and detailed information about the islands was required. In 1829 the Duchy sent their surveyor, Edward Driver, to make a comprehensive survey and valuation. He calculated that the total acreage was 3608, of which 1809 acres—more than the present figure—was in cultivation. He was concerned about the many tiny holdings which were often too small to support the tenants and their families. These arose from the system of land tenure whereby already small holdings were subdivided between the sons of the tenants on their death so that some held less than an acre, others two or three, and few more than four acres. Although the steps taken by Smith did a great deal towards consolidating small holdings, this system was partly responsible for the characteristic field pattern which was later to prove so suitable for the bulb industry. In 1831 George Driver, the brother of Edward, took possession of the islands on behalf of the Duchy and prepared a further report. The maps produced by their joint effort are still at Tresco Abbey (with a photocopy at the British Museum) and are detailed and include the uninhabited islands. They are of great value in showing the enclosures at the time. Some of these enclosures made by a land-hungry people under great economic stress have since gone out of cultivation. Without the information provided by these maps it would be difficult to interpret the vegetation on some of this land today.

Woodley in 1822 records the chief crops as wheat, barley, rye, pillas and potatoes. Pillas were a primitive oat (*Avena nuda*, see p 300)

of which the cultivation survived in Cornwall and Scilly long after
it had been superseded by other oats elsewhere in England. Potatoes
had been cultivated in quantity since before 1744 (Heath), were
exported in big tonnages during the Napoleonic Wars, and have
been an important crop ever since. Advantage was taken of the
climate to grow early varieties which were dug in April or even
earlier and the ground left fallow in the summer would be open to
immigrant weeds. Agriculture was organised on the basis that,
instead of enclosing animals as is the practice now, animals were
allowed to run free and crops enclosed to exclude them. There were
large numbers of small sheep, described as goat-like, and similar
to breeds seen in the Western Isles of Scotland, which lived mainly
on the shores on algae and fescues. As pigs were kept similarly the
pressure on native vegetation must have been very great indeed.

  The report of Scott and Rivington in 1870 described the agriculture
after Augustus Smith's reforms had begun to take effect, and the
farmers were said to be prosperous. The fields were still small.
On St Mary's they ranged from one rood to three acres, averaging
about an acre, but on the off-islands they were yet smaller. Potatoes
were the main crop but wheat and barley were grown in considerable
quantity, and on St Martin's they grew some rye. Clover was grown
extensively and mixtures of red and white clover with Black Medick
(*Medicago lupulina*) and Italian Rye-grass (*Lolium multiflorum*) were
generally sown on barley land. Crimson Clover (*Trifolium incarnatum*)
was frequently sown, and mangel and turnip also grown. These
practices throw light on differences between the observations made
by Townsend on his visit in 1862 and modern occurrences of agricultural weeds. By the time of Brewer's report on *Market-gardening
in the Scilly Islands*, 1890, the cultivation of flowers had become
important but they were still exporting 1,030 tons of potatoes a
year, 150 tons of other vegetables, 5 tons of tomatoes, and at least
3 tons of seakale. Soon after this the bulb industry became the main
use of land in Scilly.

  Before discussing this modern development other aspects which
have a long history must be mentioned. From the earliest times Scilly
was an important port of call or shelter for shipping. In the days of
sail there were sometimes as many as 200 vessels in the 'roads'
waiting for favourable winds, and opportunities for the introduction
of alien plants were common. After the introduction of steam, few
ships called, but the Scillonians went out to meet passing vessels to

exchange fresh produce for cash or goods. It is known that some garden plants were introduced in this way. Until stopped by the authorities, smuggling was rife, including rowing across to France for illicit trade. Wrecks through the ages, and indeed until the present time, have provided abundant opportunities for the introduction of alien species. Cargoes like grain, Indian corn, and even cattle thrown overboard to lighten stranded vessels, and seed salvaged from wrecks and planted might well bring introduced plants. In 1940 many islanders told me they were growing Arran Chief potatoes from a recent wreck.

Kelp-making was an industry which lasted from 1684 until 1835, and was sometimes carried on on a large scale. Seaweeds, mainly the brown wracks *Fucus nodosus*, *F. vesiculosus*, and *F. serratus*, were collected, dried in the sun, and then burnt in pits to produce a dark-grey mass known as kelp. From this, soda and potash were extracted and used in glass and soap manufactures at Bristol and elsewhere. Kelp pits are still to be seen (for example on White Island, St Martin's and Toll's Island, St Mary's) but I have failed to detect any special features of the vegetation. Where the industry did have an impact on the vegetation was on the smaller islands on which whole families took up residence for several months each year, and here the disturbance was considerable.

Another use of seaweed which goes back for many centuries is as manure on arable land. A good dressing was regarded as 12 tons an acre (*Scillonian* **31**, 45, 1956), and as recently as 1941 seaweed was the chief manure used in Scilly. It is still in use, and after gales great quantities are carted inland and left in heaps to dry before spreading it on the fields. Centuries of the use of seaweed on this scale must have had an influence on the soil, including that on small plots no longer cultivated. On the sites of inland heaps Sea Rocket, *Cakile maritima*, Sea Beet, *Beta maritima*, and other maritime plants grown from seed brought in with the weed, are to be seen.

TRESCO ABBEY GARDENS

The close association of the islands with shipping has long facilitated the introduction of attractive plants from overseas. As early as about 1650 the Godolphins' stewards were growing rare exotic plants in their garden on St Mary's which they had obtained from passing ships. An American Aloe which flowered in the garden of the Lieut-Governor on the Garrison is mentioned by several writers

(Forbes, 1821, 43–50; Woodley, 1822, 82). There can be little doubt that Augustus Smith was well aware of the scope for growing interesting plants when he took up his lease in 1834. He can hardly have foreseen that his plans for educating the islanders were to lead to many Scillonians obtaining their master's tickets and bringing back for him a steady supply of plants from the Southern Hemisphere.

In 1835 he commenced building the house to be known as Tresco Abbey and providing the necessary shelter from the winds for a garden. Elms, Sycamore, Oak and Poplar were planted on the lee side of the hill above the Abbey and later the shelter was improved with Holm Oak, *Quercus ilex*, *Cupressus macrocarpa*, and Monterey Pine, *Pinus radiata*. To plant the garden he obtained roots and seeds from Australia, New Zealand, South Africa and other countries, and from Kew. By 1860 the garden was famous for growing plants in the open for which hothouses were required at Kew (Lewes, 1860). When Augustus Smith died in 1872 the list of tender exotics was a long one including, for example, over fifty species of *Mesembryanthemum* (Anon, 1872).

Thomas Algernon Dorrien-Smith, his successor, took up residence in 1874 and continued to introduce new plants to Tresco. By 1898 there were 120 varieties of *Mesembryanthemum* in cultivation (Fitzherbert, 1898). When he died in 1918 he was succeeded by Major Arthur A. Dorrien-Smith, who had already collected plants in the Southern Hemisphere to be grown in Scilly (Dorrien-Smith, 1908, 1910, 1911). To his devotion to Tresco Gardens and his great knowledge of botany and horticulture we owe the marvellous collection to be seen today. As an indication of the continued growth, the garden list recorded 153 species of *Mesembryanthemum* by 1935 (Hunkin, 1947). Major Arthur Dorrien-Smith died in 1955 and was succeeded by his son Lieut-Commander Tom Dorrien-Smith, who is maintaining the famous gardens in collaboration with the Royal Botanic Gardens, Kew.

From Tresco Abbey many plants have been distributed to other parts of the islands, and some have become established in wild situations. Within the gardens many exotic species reproduce themselves freely from seed, thus demonstrating the possibility that they might persist. For example E. Brown wrote in 1935

> One of the sights which most impressed me was the great number of plants which reproduce themselves naturally from seeds. Cordylines, Palms, Myrtles, Gunneras, Olearias, Fuchsias, Pelagoniums, Gazanias,

Sempervivums, Veronicas, Echiums, and hundreds of other plants could be seen in all stages of growth beneath their parents, *Agapanthus* simply abounded, and its seedlings were to be found in woods, in paths, between rocks, and on the top of walls.

The only exotic species growing in the gardens which have been included in this *Flora* are a few with a long history of persistence as weeds, and these mainly on the special advice of Major Dorrien-Smith or of the gardeners. Exotics growing on Appletree Banks or elsewhere in Tresco are only included for similar special reasons. It was the practice of the Major to walk about with bits of succulents in his pockets to press into the ground wherever he thought they might grow—as they often did.

Most of the windbreaks to be seen in all the inhabited islands originated from Tresco Abbey. *Pittosporum crassifolium*, which came from New Zealand, is an outstanding case. This evergreen was found to resist salt spray and stand up to the wind and was widely planted. Subsequently the seeds were distributed by birds to uninhabited islands and sea-girt rocks. *Olearia traversii* was brought by the Major from Chatham Island in 1911 and comes freely from seed. *Escallonia macrantha* grown in the gardens from Chile is now widespread as a windbreak. The 'Mesems' to be seen throughout the inhabited islands, and indeed in the case of Hottentot's Fig *Carpobrotus edule*, on uninhabited ones also, probably all came from Tresco Abbey Gardens. The smaller species were first taken to adorn cottage gardens, and then pieces got pushed into walls and rocks away from houses. Although plentiful, many of them can only be regarded as garden plants, but a few species spread from seed or successfully compete with native vegetation, and have therefore found a place in this *Flora*. Tresco Gardens have done much to change the face of Scilly.

THE FLOWER INDUSTRY

The Scillonian Flower Industry, which became the main support of the economy of the islands, has grown up during the last hundred years. It started with a single box of cut narcissi sent to Covent Garden market in 1865. (Accounts of the early history of the industry differ considerably, and this date has been given as 1867, 1870 or even 1875. I follow Dorrien-Smith (1890) who was in the best position to get it right.) At first it was confined to a few farms—the Abbey Gardens in Tresco, and farms at Rocky Hill, Holy Vale,

Old Town and Newford in St Mary's. After 1883, when T. A. Dorrien-Smith visited Belgium, Holland and the Channel Isles to study their methods and import new bulbs, the industry made rapid growth (Table 4).

Table 4   GROWTH OF THE FLOWER INDUSTRY

The following figures show the tonnage exported in each year. They have been obtained from various sources and are not strictly comparable owing to changes in the methods of packing.

	Tons		Tons
1885	65	1896	514
1886	85	1901	650
1887	100	1904	800
1888	188	1924	700
1889	198	1931	1061

The narcissi first grown were found in a quasi-wild state in the islands. Whatever their origin may have been, they had multiplied so freely that by the middle of the nineteenth century the bulbs had become an agricultural pest on Tresco and St Mary's. They are best described in the words of T. A. Dorrien-Smith (1890):

'Twenty-five years ago (1865) some eight varieties of the Narcissi were growing in the Isles of Scilly, besides those in the Abbey gardens, some almost wild, some in the hedges, and some in the gardens attached to the little farms.

These varieties were: *Telamonius plenus, odorus major* (Campernelli), *Tazetta ochroleucus* (Scilly White), *Tazetta aureus* (Grand Soleil d'Or), *Tazetta* Grande Monarque (two vars.), *biflorus, poeticus fl.pl.*, and *poeticus recurvus*.

The date of the introduction of these is extremely obscure, except Campernelli of which two bulbs were presented to Mrs Gluyas fifty years ago by the captain of a French vessel, and her son now holds the largest stock. The others were introduced probably by the Governors, the largest stock being found around their country seat at Holy Vale, St Mary's, or they may have been introduced by the monks resident at Tresco (these monks belonged to the Abbey of Tavistock, and were of the Benedictine order). There were most of the above varieties growing in the Abbey gardens in the vicinity of the old Abbey ruins, as well as many varieties which had been introduced since 1834 by Mr Augustus Smith. . . .'

There is some support for the suggestion that Scilly White was introduced by the monks of the Benedictine order in that this narcissus is said to have grown 'wild' from time immemorial on St Michael's Mount near Penzance, and Mont St Michel in Brittany. Until the reign of Henry IV the monks of St Michael's Mount and Tresco belonged to the Benedictine order, of which Mont St Michel was the seat of the original foundation (Taylor, 1906). On the other hand, Gray (1933) regarded Scilly White as a collected stock of *Tazetta ochroleucus* or *T. laeticola* from the south of France unlikely to have been in Scilly before the end of the eighteenth century. However that may be, Scilly White and the 'Old English Daffodil' appear to have been thoroughly established and on St Mary's they grew in abundance at Holy Vale, Newford, Trenoweth and elsewhere. Soleil d'Or, *Narcissus aureus* Lois., now the mainstay of the Christmas flower trade, is a native of southern France and Italy. It has been known in England since the days of James I, but it is only in the mild winters of Scilly that it thrives in the open. Even in the Channel Isles it is almost wiped out by harsh winters.

Since the early days of the industry many new bulbs have been tried and some have gone into commercial production, but the success of the industry has always depended on the early flowering kinds. It is only by exploiting their winter climatic advantages to forestall growers on the mainland that the Scillonians can overcome the handicap of heavy freight charges involved in the sea crossing in addition to the long rail journeys from Penzance to the markets. The flower season now opens by sending Paper White to market in November, closely followed by Soleil d'Or and Rijnveld's Early Sensation (Ward, 1964). It continues with the first daffodils, Magnificent and Forerunner, towards the end of January, and reaches a hectic peak in the third week of March with Cheerfulness and Late Double White. In mid-April it closes with tulips and iris. That is in a normal season, when the flowers will be picked in bud and there will be little colour in the fields. In a winter with unfavourable weather the bulbs will be retarded and mature with a rush. Then the inability of pickers to deal with so many blooms, and glutted markets, may leave bulb squares unpicked and a sheet of colour. In March 1963, which followed a severe winter, the flowers came on in such quantity that a record 14,000 boxes, about 70 tons, were despatched in a single shipment and even then the fields were left full of bloom.

Ideally the bulbs remain in the ground for three years, though some are better lifted after two, and Soleil d'Or can be left without harm as long as five. Lifting goes on from May to August, and replanting in June for Soleil d'Or, and from August onwards for others. In the early days of the industry there was a rotation with mangolds or potatoes, and grass, but bulbs were so much more profitable that farmers were tempted to plant them in more than a third of their arable, and there was a great increase in disease. The present tendency is to use Rye-grass and other leys, and there is less dependence on seaweed as a manure.

In the bulbfields weeds remain undisturbed throughout the winter months and thorough cultivation is possible in only, on an average, one year in three. This has facilitated a very special type of weed flora characterised by species which grow through the winter, flower in spring and early summer, and tend to have their resting period in summer when the shallow light soils of Scilly are often dry. This life-cycle is that of Mediterranean annuals, and most of the species concerned have a Mediterranean type of distribution such as: *Ranunculus muricatus*, *R. marginatus*, *Fumaria occidentalis*, *Silene gallica*, *Polycarpon tetraphyllum*, *Medicago polymorpha*, *M. arabica*, *Vulpia bromoides*, *Briza minor* and *Bromus diandrus*, or from a country with a similar climate, like *Crassula decumbens*. This type of cultivation also makes it extremely difficult to eradicate bulbous weeds such as *Oxalis pes-caprae*, *Ornithogalum umbellatum*, *Endymion hispanicus* and *Allium roseum* subsp *bulbiferum*.

The conditions which suit Mediterranean species have also been brought about by changes in microclimate resulting from the bulb industry. Protection from the wind has always been essential for cultivation in Scilly. This was first provided by granite boulder walls known as 'hedges'. Often about a metre wide, and earth-filled, these had a flora of their own like their counterparts in Cornwall. Normally each field, as shown on the Ordnance Survey maps, is surrounded by a hedge and when used for bulbs is divided into units known as 'squares' by a series of windbreaks called 'fences' (Plate p 161). These fences are made up of evergreen shrubs of species selected for rapid growth and ability to withstand gales and salt spray. The shrubs in general use are *Pittosporum crassifolium*, *Excallonia macrantha*, *Hebe lewisii*, *Brachyglottis repanda* and *Euonymus japonicus* (Fig. 9).

Prior to the arrival of Augustus Smith, attempts had been made to provide windbreaks by planting elms and tamarisk, and these still occur as relics. But elms give no protection in winter when it is most needed, and tamarisk offers little resistance to wind. Experience on

Tresco led to the use of the *Escallonia* and *Hebe* together with *Euonymus japonicus* in the vegetable-growing days of the sixties, and they were readily available when the flower industry started. Later *Pittosporum* was added to provide higher fences. It takes five years to grow a fence 15 ft tall, and it will last for thirty years or more. Bulb squares surrounded by hedges and fences develop temperatures significantly higher than outside. An account of their weeds is given later (pp 76–78).

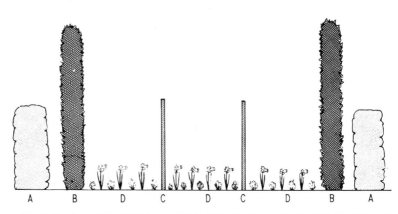

Fig 4 Section of Bulbfield: A. 'Hedge' (wall of granite blocks). B. 'Fence' (row of close evergreen shrubs). C. Movable screens. D. Narcissi with abundant over-wintering weeds. (Adapted from Leuze, 1966.)

MODERN FARMING ECONOMY

The pattern of agriculture in Scilly is best illustrated by the annual statistics of the Ministry of Agriculture, Fisheries and Food. A summary of the returns for June 1967 (the last available) is given in Table 5.

Out of a total area of 3,980 acres less than half were cultivated. Arable totalled 919 acres, permanent grass 312, and 331 were used as rough grazing. Out of the arable, 536 acres were used for flowers and bulbs grown in the open and much of the fallow and clover and rotation grasses would be land being rested from bulbs. Other uses of the arable are shown in the table and it will be seen that no cereals were grown; even oats, of which there were a few acres right up to 1966, are no longer included. This explains the decrease in cornfield weeds recorded by earlier botanists. On the other hand

Table 5. AGRICULTURAL STATISTICS FOR ISLES OF SCILLY AS AT JUNE 1967

Adapted from Ministry of Agriculture & Food, Agricultural Statistics 1967/68. England and Wales. 1970

	Acres	
Crops and fallow (excluding lucerne). (Tillage)		
Potatoes, first earlies	48	
Potatoes, maincrop and second earlies	28	
Turnips, swedes, fodder beet, cabbage, savoy, kohl rabi, kale, etc. for stock feeding	49	
Mangolds	12	
Rape	2	
Mustard for seed, fodder or ploughing in	4	
Orchards not grown commercially	1	
Vegetables (excluding potatoes) grown in the open	13	
Other crops	1	
Bulbs and flowers grown in the open and nursery stock	536	
Bare fallow	72	
		766
Lucerne	1	1
Clover and Rotation Grasses		
For mowing	58	
For grazing	94	
		152
Permanent Grass		
For mowing	85	
For grazing	227	
		312
TOTAL CROPS AND GRASS		1231
Rough Grazings (Sole rights)	331	331
TOTAL		1562

(The total area (excluding water) of the Isles of Scilly for the purposes of these returns is 3980 acres).

the sowing of temporary leys, which at 152 acres are rather less than most recent years, explains the introduction of grasses such as Timothy, Italian Rye and Meadow Foxtail and other plants which are now recorded. Another trend is for a reduction in arable owing to land going out of cultivation on the off-islands. In part this arises from increasing difficulties and rising costs in transporting flowers to the quay at St Mary's, in part to falling population as younger people move away.

Repeated attempts have been made to drain Higher and Lower Moors, St Mary's in the interests of agriculture but with significant lack of success. More serious damage to the flora has arisen from the withdrawal of grazing and especially from the part of Higher Moors north of the road between Tremelethen and London. Coarse grasses have taken over these fields and the only bog in Scilly is almost lost. On the other side of the road attempts were made in 1954 to destroy with toxic chemicals the wonderful tussocks of the sedge *Carex paniculata* with their epiphytic flora (see p 285). The spray greatly reduced the Royal Fern and other epiphytes but there has been some recovery. Herbicides are used fairly heavily on agricultural land and occasionally misused by spraying roadside hedges for which there seems no justification.

Mechanisation led to the number of horses falling rapidly from 160 in 1943, to 92 in 1955 and now to very few. As a result there are considerable changes due to withdrawal of grazing as, for example in Lower Moors, where in 1967 there was just one old horse and much coarse vegetation, and on Wingletang Downs, St Agnes where gorse has greatly increased. Similarly there has been a big fall in the the number of donkeys since I first knew Scilly and the small roadside patches where they were formerly tethered are now overgrown.

PEAT CUTTING

Fuel has always been a problem in Scilly from the time the last of the trees of prehistoric times was destroyed. Gorse was one source of fuel in the Roman era (Ashbee, 1955) and no doubt has been an important source for some purposes since, but the main fuel in Scilly until recently was peat.

This was mainly peat derived from Ling, *Calluna vulgaris*, on the higher ground and exposed slopes. Woodley in 1822 records that Dolphin Downs on Tresco were much injured by being pared for fuel, and turf was dug out on Northern Hill on Bryher. Both places still show evidence of this. Turfy Hill on St Martin's was so called from the quantity of turf cut (North, 1850, 63). On St Mary's peat was still cut from the Downs almost within living memory and Trevellick Moyle remembered that the Rocky Hill Farm plot was on Giant's Castle Down. They carted it home and stored it for winter use.

About 1695 a sort of bituminous earth was dug in plenty in the moors in St Mary's and Tresco. In digging the peat they found trees two to eight ft in circumference at a depth of five to six ft (Turner, 1964). The nature of this peat is not known but Mr P. Z. MacKenzie recently found peat at Porthellick formed from willow, birch, sedges, grasses, etc (Dimbleby, ined.). While there is no proof, the possibility must be borne in mind that the open water at Porthellick Pool, and at Great Pool, Tresco, may be in part ancient peat workings which have been flooded.

TOURIST INDUSTRY

During the last twenty years there has been outstanding growth in the number of visitors to the islands, so that tourism is replacing flower growing as the major industry. By 1959 the number of staying holiday-makers averaged 13,000 annually and the impact on the vegetation of the six and a half square miles is considerable.

In order to accommodate the additional visitors and others, new buildings have been necessary, and it is fortunate that the enlightened policy of the Council has almost restricted this to infilling and fringe extension of Hugh Town and Old Town on St Mary's. Small pieces of waste ground which were the homes of plants such as White Mignonette, *Reseda alba*, and Sweet Alyssum, *Lobularia maritima*, known for over a century, have almost gone. The inevitable tidying-up which always takes place in tourist haunts has recently almost destroyed the botanical interest of Porth Cressa, where Sea Radish, *Raphanus maritimus*, and other maritime plants formerly had such a fine display. Fortunately there has been very little building on the other islands apart from the new hotel on Tresco.

A more serious aspect is the impact on the vegetation from trampling and fires. Even a few days of fine weather are commonly followed by gorse and heath fires and after a drought these reach serious proportions. For example, in 1949 fires on Bryher and St Helen's burnt for days, and picknickers caused another large fire near Innisidgen on St Mary's. In 1955, when there were eight weeks of drought, a gorse and heath fire below the Golf Course on St Mary's set the peat alight to a depth of six inches and seven acres of down smouldered for weeks. In 1956 there was another big fire at Innisidgen, and in 1959 serious fires on Samson Hill, Bryher continued for weeks, and there were several bad fires on St Martin's. Conflagrations on this scale result in considerable changes in the vegetation since not only is the long accumulation of peat destroyed, but the soil is laid open to erosion by the winter gales.

It is fortunate that the uninhabited islands have protection from tourist pressure. Landing on Annet, the bird sanctuary, is strictly controlled, and on some others during the nesting season. Also the very volume of the visitors brings in itself some protection to the smallest islands since the boatmen have no time to go out of their way to ferry the crowds ashore. But Samson and Tean suffer considerably, and the impact on some of the others will be referred to later in this book.

Sufficient has been said to show that it is impossible to interpret the vegetation without knowledge of the influence of man throughout history, or to understand recent changes in the recorded plants without reference to changes in land use.

Chapter Three

GEOGRAPHICAL RELATIONSHIPS

There are two aspects of the flora which an observant visitor to Scilly can hardly fail to notice. Many of the common and conspicuous plants he has seen travelling down through Cornwall by train or road do not occur and others are local or scarce. Also on the uninhabited islands he will be impressed with how few species occur, far less than on the larger ones. This poverty in numbers is characteristic of island floras, but what they lack in floristic richness is more than made up by the interest, and often abundance, of the plants which do occur.

COMPARISON WITH THE CORNISH MAINLAND

At Land's End, the nearest point on the mainland to Scilly, there are species growing in abundance such as Kidney Vetch, *Anthyllis vulneraria*, and Heath Pearlwort, *Sagina subulata*, which do not occur in the islands. These are mixed with flowers which are abundant in Scilly, like Thrift, *Armeria maritima*, and Sea Pearlwort, *Sagina maritima*, and others such as Vernal Squill, *Scilla verna*, which are here in far greater quantity. It is this change in emphasis which is more impressive than simple statements of presence or absence. It is the abundance of Greater Stitchwort, *Stellaria holostea*, and Herb Robert, *Geranium robertianum*, about Penzance which emphasises that failure to find them in Scilly must have some scientific explanation.

The differences in climate between Scilly and Penzance discussed on page 25 explain why winter growing plants have an advantage in the islands, but show no advantages to mainland species, even in rainfall, to account for the greater numbers in Cornwall. It is tempting to seek a historical explanation and to postulate that some of the plants which have entered Britain in successive waves from the continent have reached west Cornwall and failed to cross the sea from there to Scilly, just as others have failed to cross the Irish Sea to Ireland. This is true of some species, but fossil records show that most of those concerned have been in Britain for a very

Table 6 SPECIES FREQUENT IN CORNWALL BUT NOT PRESENT IN SCILLY

(Based on the *Atlas of the British Flora*. The list covers species which occur in at least ten km squares in Cornwall and are regarded as absent from Scilly. For those marked with an * there are records from the islands which have not been confirmed or which refer to garden plants.)

WOODLAND PLANTS

Anemone nemorosa, *Clematis vitalba*, *Alliaria petiolata*, *Hypericum androsaemum*, *Stellaria holostea*, *Moehringia trinervia*, *Oxalis acetosella*, *Euonymus europaeus*, *Potentilla sterilis*, *Fragaria vesca*, *Geum urbanum*, *Agrimonia eupatorium*, *Rosa arvensis*, *Sorbus aucuparia*, *Sanicula europaea*, *Mercurialis perennis*, *Ulmus carpinifolia*, *Betula pendula*, *B. pubescens*, *Fagus sylvatica*, *Corylus avellana*, *Quercus petraea*, *Lysimachia nemorum*, *Fraxinus excelsior*, *Melampyrum pratense*, *Betonica officinalis*, *Allium ursinum*, *Tamus communis*, *Orchis mascula*, *Arum maculatum*, *Holcus mollis*.

BOG PLANTS

Viola palustris, *Drosera rotundifolia*, *Scutellaria minor*, *Narthecium ossifragum*.

HEATH PLANTS

Viola lactea, *Sagina subulata*, *Illecebrum verticillatum*, *Salix aurita*, *S. repens*, *Erica tetralix*, *E. vagans*, *Vaccinium myrtillus*, *Achillea ptarmica*, *Carex ovalis*, *Deschampsia cespitosa*, *Agrostis setacea*.

MARITIME PLANTS

Salicornia spp., *Apium graveolens*, *Bromus ferronii*.

PLANTS OF WET PLACES

Equisetum palustre, *Caltha palustris*, *Hypericum tetrapterum*, *Chrysosplenium oppositifolium*, *Symphytum officinale*, *Scrophularia aquatica*, *Veronica beccabunga*, *Senecio aquaticus*, *Eupatorium cannabinum*.

PLANTS OF VARIOUS HABITATS

Asplenium ruta-muraria, *Aquilegia vulgaris*, *Reseda luteola*, *Viola odorata*, *Hypericum perforatum*, *Silene vulgaris*, *Geranium columbinum*, *G. robertianum*, *Trifolium medium*, *Anthyllis vulneraria*, *Vicia sepium*, *Poterium sanguisorba*, *Chaerophyllum temulentum*, *Primula veris*, *Rhinanthus minor*, *Odontites verna*, *Mentha arvensis*, *Euphrasia nemorosa*, *Knautia arvensis*, *Succisa pratensis*, *Narcissus pseudo-narcissus* (as a native), *Phleum pratense* (not cultivated).

COMMON ALIENS WHICH HAVE NOT REACHED SCILLY

Cardaria draba, *Epilobium angustifolium*, *Castanea sativa*, *Buddleja davidii*, *Lamium album*, *Mentha x piperita*.

long time and it is hard to believe that a sea barrier of only thirty miles could prove so effective. Fortunately there is a very much simpler explanation.

Islands, all over the world, are notably poor in species when contrasted with mainland areas of similar size, climate and diversity. In the case of small islands the number of species tends to be roughly proportionate to the area—the smaller the island the fewer species represented. As a broad generalisation this must be true because the coast is subject to the immediate impact of gales and salt spray, the influence of which falls off with shelter and distance from the sea. Thus, even in St Mary's, which with 1,562 acres is the largest island in Scilly, the range of habitats is limited. There is no real woodland, only a minute bog, no sizable stream, only limited freshwater habitats, and no salt marsh. Table 6 shows species frequent in Cornwall which do not occur in Scilly and most of them are plants one would not expect to occur in the habitats available. Flowers will grow only where conditions are suitable.

COMPARISON BETWEEN THE ISLANDS

This is confirmed by a comparison between the islands of the archipelago. As will be seen from Table 7 the number of species in each of the five large islands is related to their size. St Mary's with

Table 7 NUMBER OF SPECIES IN RELATION TO AREA

	Acreage above HWM	Number of Recorded species
INHABITED ISLANDS		
St Mary's	1562	582
Tresco	735	473
St Martin's	589	329
St Agnes	365	323
Bryher	327	285
UNINHABITED ISLANDS		
Samson	102	127
Annet Group	84	53
Eastern Isles	79	111
St Helen's Group	74	97
Tean	44	119

1,562 acres and Tresco with 735 are sufficiently large to have lakes and marshes with fresh, or almost fresh, water. These and other habitats are sheltered from the extreme maritime influence, and their size provides considerable diversity and large cultivated areas with arable weeds. St Martin's is much less diversified, there are no lakes or marshes, much of the island is sandy, and the rest heathland or coastal. St Agnes, though smaller, is a little more diversified with its pool and some moisture-loving species but, this, as in Bryher, does not make up for lack of size. This can perhaps be shown more clearly by a comparison which omits the commonest plants. There are 273 species found in all the inhabited islands, or recorded from all but one. Deducting these, there are over three hundred additional species on St Mary's, two hundred on Tresco, about fifty on St Martin's and St Agnes, and only twenty on Bryher.

The figures for the uninhabited islands are more difficult to interpret. As will be shown in the next chapter, Samson and Tean owe their high totals to relatively recent occupation and cultivation. The Eastern Isles are comparatively sheltered and the St Helen's group includes high ground, but Annet is so exposed to salt spray that only a limited number of species can persist.

COMPARISON WITH THE CHANNEL ISLES

The Isles of Scilly have much in common with the Channel Isles. The scenery of both is based on exposures of granite linked by sand and other granite-derived deposits, and surrounded by sea. The climates, as already shown on p 25, are similar, although extreme winter temperatures do not fall as low in Scilly as they do in Jersey. Both take advantage of their mild climates to grow tender plants which will not thrive on the mainland, and both grow early bulbs with similar weeds.

It is therefore not surprising that they have so much in common in their native vegetation and especially in that of the coast. On the island of Herm, for example, the botanist might almost imagine himself back in Scilly: the communities are almost identical so far as ecologically important plants are concerned and there are only a few differences in the floristic lists. There are much closer resemblances between the floras of Scilly and the Channel Isles than there are between Scilly and the nearby mainland of Cornwall.

Species found in Scilly and/or Cornwall which are very rare in the British Isles are listed in Table 8 with a statement of their status in the Channel Isles. The following depend on the winter and early spring for growth and reproduction, that is, they have a life cycle suited to a Mediterranean type of climate: *Isoetes histrix*, *Ophioglossum lusitanicum*, *Viola kitaibeliana*, *Spergularia bocconii*, *Polycarpon tetraphyllum*, *Lavatera cretica*, *Ornithopus pinnatus*, *Arum italicum*, *Juncus capitatus*, and *Poa infirma*. The following are

Table 8   RARE SPECIES OF SCILLY AND CORNWALL WITH THEIR STATUS IN THE CHANNEL ISLANDS

	Scilly	*Cornwall*	*Channel Islands*
Isoetes histrix	absent	very rare	less rare
Adiantum capillus-veneris	very rare	rare	rare
Ophioglossum lusitanicum	very rare	absent	very rare
Viola kitaibeliana	rare	absent	local
Spergularia bocconii	very rare	very rare	local
Polycarpon tetraphyllum	abundant	very rare	abundant
Lavatera cretica	common	very rare	frequent
Trifolium occidentale	frequent	rare	frequent
T. molinerii	absent	very rare	very rare
T. bocconei	absent	rare	very rare
T. strictum	absent	rare	very rare
Lotus hispidus	common	rare	frequent
Ornithopus pinnatus	common	absent	rare
Euphorbia peplis	extinct	extinct	very rare
Polygonum maritimum	extinct	? extinct	very rare
Rumex rupestris	common	rare	rare
Echium lycopsis	very rare	local	local
Scrophularia scorodonia	abundant	frequent	frequent
Otanthus maritimus	extinct	extinct	extinct
Allium babingtonii	common	local	absent
Juncus capitatus	? absent	local	local
Arum italicum	common	frequent	local
Briza minor	abundant	frequent	scarce
Poa infirma	common	local	common

very frost-sensitive: *Adiantum capillus-veneris, Otanthus maritimus,* and the following are limited to climates with very mild winters: *Trifolium occidentale, T. molinerii, T. bocconei, T. strictum, Lotus hispidus, Echium lycopsis, Scrophularia scorodonia, Allium babingtonii* and *Briza minor*. It is evident that all these plants grow where they do because the Channel Isles, Scilly and Cornwall alone in Britain provide the mild winters they require. These, together with the extreme west of Ireland, have over 350 growing days in the year, that is, with average air temperatures in excess of 5°C. The other three plants in the list are restricted to shingly and rocky shores.

There are also close resemblances in the introduced species which have become established and are now part of the flora. The following are examples: *Portulaca oleracea, Carpobrotus edulis* and other 'Mesems', *Oxalis articulata, O. pes-caprae, Muehlenbeckia complexa, Nicandra physaloides, Solanum sarrachoides, Senecio mikanioides,*

Endymion hispanicus, Muscari comosum, Narcissus spp., Bromus willdenowii, B. diandrus, B. madritensis.

MIGRATION THEORIES

It has been shown that the main reason why many of the plants common on the mainland fail to reach Scilly is that suitable habitats are not available, and that the insular climate enables winter-growing plants to thrive. These two factors account for most of the peculiarities of the flora of Scilly. The thirty miles of sea separating the islands from Cornwall have no doubt acted as a barrier to some species, but it is unrealistic to build up elaborate migration theories.

Oceanic islands have outstanding biological characteristics which have attracted the attention of many leading workers. Wace and Dickson (1965) have provided an excellent summary of the characteristics of island floras, with a good bibliography, in their recent analysis of the botany of the Tristan da Cunha group of islands and reference should be made to this. In such cases isolation acts as a real barrier and the floras exhibit the classical features of island life, such as endemism, ecological disharmony, species poverty and radiation. Off-shore islands like those of Scilly show these in only a very minor degree. Of endemic plants it is possible only to point to the doubtful case of the rush, *Juncus maritimus* var *atlanticus*, in mammals to the Scilly Shrew, *Crocidura suaveolens cassiteridum*, in butterflies to some slight variations of the Meadow Brown and Common Blue. Thus there is no evolutionary radiation, and I have found no evidence of ecological disharmony, but as already shown, there is species poverty. Even this is less than might be expected. Clare Island, off the coast of West Mayo, with about the same area (4,025 acres), similar maritime influence and humid atmosphere, but with greater ecological diversity permitting the presence of alpines, bog and salt marsh plants, has only 393 species compared with the 687 of Scilly (Praeger, 1911).

J. C. Willis used the Isles of Scilly to support a side issue of his theory in his book *Age and Area* (1922). He postulated that in islands lying off a mainland the flora will be made up mainly of representatives of the larger families and genera in the country, that is, those, which according to his theory, had been in existence longest. This was based on his observation that species with a wide range generally tend to form a high proportion of the flora of islands. There was no need to invent an elaborate theory to explain this. The flora of every part of the British Isles includes a long list of adaptable successful plants like Bracken, Tormentil, Ivy, Ling, Honeysuckle and Foxglove, and weeds of cultivation such as Shepherd's Purse, Chickweed, and Annual Meadow-grass. Obviously if there are fewer species present, these will form a higher proportion of the flora.

H. B. Guppy (1925) challenged Willis' work. He found that the plants of Scilly which occurred on the Cornish mainland had a much greater average range in the county than the Cornish plants which were not recorded from the islands. Looking into it more closely he 'found in those islands a gathering of plants that in many cases had long since proved their capacities for extending their ranges by travelling over much of the globe', cosmopolitan plants, many of them weeds of cultivation. 'It is a remarkable fact', he wrote, 'that the flora of the Scilly Islands is much more Azorean than that of the Cornish peninsula, almost half (49 per cent) of the Scillies being Azorean species, whilst only one-seventh of the purely Cornish plants are held in common with the Azores'. He goes on to explain that these colonising cosmopolitan plants have largely displaced the native flora of the Azores. Guppy's work brought to light interesting facts, but it seems to me that resemblances in climate, and suitable habitats, are the main explanations.

COMPARISON WITH EUROPEAN DISTRIBUTION

It will be evident that statistics including a high proportion of cosmopolitan species are unlikely to throw much light on the origin of the flora of Scilly. A very much more promising approach is to consider the less generally widespread plants in relation to their total European distribution. This is the approach adopted by Professor J. R. Matthews in his works on the distribution of the British flora (1937, 1955). Excluding very widely distributed plants and endemics, he divided the British flora into twelve 'elements' which are represented in Scilly as shown in Table 9.

The Mediterranean element is exceptionally well represented with 42 per cent of the species occurring in Britain. These are plants whose chief centre of distribution is the Mediterranean region, with extensions through western France. They include *Bromus madritensis*, *Echium lycopsis*, *Euphorbia peplis*, *Geranium purpureum*, *Lavatera arborea*, *L. cretica*, *Otanthus maritimus*, *Poa infirma*, *Polygonum maritimum* and *Spergularia bocconii*. The Oceanic Southern element

Table 9   THE GEOGRAPHICAL RELATIONSHIPS OF THE FLORA OF SCILLY

Based on the 'Geographical Elements in the British flora' as listed in Matthews, 1955

Element	Number of Species		
	in Britain	in Scilly	% in Scilly
Mediterranean	38	16	42%
Oceanic Southern	82	32	39%
Oceanic West European	87	30	34%
Continental Southern	130	31	24%
Continental	88	5	6%
Continental Northern	97	6	6%
Northern Montane	31	—	—
Oceanic Northern	23	8	35%
North American	6	—	—
Arctic-Subarctic	28	—	—
Arctic-Alpine	75	—	—
Alpine	10	—	—

is represented by 39 per cent of the British species. This is essentially a south-west European group of species which occur chiefly in southern Europe, including the Mediterranean region, and in western Europe. Of the thirty-two plants of this group in Scilly, *Ornithopus pinnatus*, various clovers (*Trifolium glomeratum, micranthum, subterraneum, suffocatum* and *ornithopodioides*) and *Umbilicus rupestris* are examples. The Oceanic Northern element is well represented with 35 per cent of the British species. This is essentially a North Atlantic assemblage and all but one of those in Scilly are plants of seashores. The Oceanic West European are plants found almost exclusively in western Europe—such as *Anthemis nobilis*, *Erodium glutinosum* and *E. maritimum*, *Rumex rupestris*, *Spergularia rupicola* and *Ulex gallii*. The other large element is Continental Southern, plants occurring in central and southern Europe but thinning out northwards. Of these we have 24 per cent of those in Britain.

The interesting feature that this reveals is the high proportion of species growing in Scilly which have spread up from the Mediterranean—forty-eight in the Mediterranean and Oceanic Southern groups. Most of these (and others) occur also in the Channel Isles. They are mainly plants of maritime shores or dry sandy soils and many depend on growth during the winter months, which is characteristic of the Mediterranean flora. It may well be that their path of migration to Scilly followed the same route round the Atlantic coast as the Bronze Age men who built the megalithic tombs in Scilly and the Channel Isles. Any plants they brought as agricultural weeds would be likely to find conditions suitable for persistence.

Very little work has been published on the geographical relationships of the flora and fauna of Scilly but this confirms the conclusions based on flowering plants. Ranwell (1966) in a careful analysis of the lichens found on Bryher shows that most of the species found have distinctively western or maritime trends in Britain. High humidity, relative freedom from frost, and the drenching from sea spray are the main ecological conditions which determine their presence.

Blair (1931) discussed the beetles of Scilly and showed that they confirmed the rule that insect life on an island is poorer both in species and individuals than the adjoining mainland. The 574 species listed include some known only in Britain from Scilly but these Dr Blair thought would be found elsewhere in England. Some belong to Scharff's Lusitanian or Atlantic element, of species with a western distribution in Europe, and others to the Germanic element, which arrived in Britain from the continent subsequent to the final retreat of the glaciers. He concludes that the absence of some of these which are common in Devon and Cornwall is due to unfavourable climatic conditions and the absence of trees and shrubs. As Blair (1925) points out, the bleak windswept conditions of Scilly are not favourable to Lepidoptera, but nevertheless the islands lend themselves to the development of minor variations through isolation. Examples are the Meadow Brown butterfly described by Graves (1930) as *Maniola jurtina* L. subsp *cassiteridum*, and the variations of the Common Blue on small islands (Ford, 1967).

Bristowe (1929, 1935) collected spiders from twenty-eight islands in three visits and has compared his results with Lundy and the Channel Isles. He found that diversity of habitats resulted in the number of species being roughly related to the size of the individual islands, but in the case of the smaller ones exposure to wind and big seas decided the plant life and hence the shelter for spiders. His total was 119 species for Scilly compared with 241 in the Channel Isles and 75 on Lundy (including Harvest Spiders and Pseudoscorpions). He found it difficult to say whether the absence from Scilly of 14 species he found near Penzance was due to failure to reach the islands or to unsuitability of climate. As with the beetles, there are a number of spiders introduced by man.

With both plants and animals it is the unusual climatic conditions and human activities which explain their presence. Most of the species which have failed to cross from the mainland would not thrive in Scilly and it is unnecessary to invoke elaborate theories of past migrations to explain this.

Chapter Four

THE UNINHABITED ISLANDS

There are few greater joys than to be set down for the day with food and drink on an uninhabited island with the certainty that the boatman will collect you in time for dinner. The sea sets a boundary to your kingdom, and the small size makes detailed and leisurely search possible. In good weather and at appropriate times of the year the brilliance of colour is almost unbelievable with Thrift, English Stonecrop, Foxglove, or other flowers set off against blue sky and sea. It is easy to suppose that this is a habitat made by the sea and the climate and the birds in which man has hardly interfered. In the case of a few of the very small islands this could be true.

WESTERN ISLES

It is almost true of all the islands I have grouped with Annet—the Western Rocks, Mincarlo, Castle Bryher, Illiswilgig, Maiden Bower and Scilly Rock (see map p 4). These are all exposed to the full force of the Atlantic and the south-west gales. At times the sea breaks right over all these islands, at others they are just soused in salt spray. No plant can grow there unless it is firmly rooted and able to thrive where the only fresh water comes as rain almost immediately made salt by the sea. On some islets such as Maiden Bower and Great Crebawethan there are no flowering plants, and these Leland, *circa* 1535, aptly termed 'blynd rokkettes'. On others almost equally exposed the only flowering plants are Thrift, Sea Beet, Common Scurvy Grass, Tree Mallow, Rock Spurrey and Orache. W. S. Bristowe, who has landed on more of these islets than any other naturalist, listed these from exposed islands in this group as follows (Bristowe, 1935):

Armeria maritima—Illiswilgig

Beta maritima—Illiswilgig, Rosevear, Castle Bryher

Cochlearia officinalis—Rosevear, Castle Bryher, Gorregan

Lavatera arborea—Illiswilgig, Rosevear, Castle Bryher

Spergularia rupicola—Illiswilgig, Rosevear, Castle Bryher, Scilly Rock

Atriplex spp—Illiswilgig, Rosevear, Castle Bryher, Gorregan, Scilly Rock

These six (several species of *Atriplex* are involved) may well hold the record for ability to withstand salt spray, but others run them very close.

Mincarlo, a sizeable island of four and a half acres, is representative of the poverty of the flora under these conditions. My most recent report is from M. Walpole, who landed there in 1967 and found *Lavatera arborea* (abundant), *Atriplex* spp (in quantity), *Cochlearia officinalis* (very local, only three or four robust plants) and *Spergularia rupicola* (two plants only). The only addition to this from earlier lists is *Beta maritima*. Rosevear is about the same size, and workmen lived there between 1847 and 1850 when the first iron lighthouse was being erected on the Bishop Rock and again when the present granite lighthouse was erected. Buildings were constructed and they even managed to grow vegetables (Williams, 1900) but did nothing to enrich the flora permanently. Only the five species listed by Bristowe (see above) are known from Rosevear today.

In fact only 53 species are recorded for the whole group of Western and Norrad Rocks and Annet—and they all occur on Annet. This is a low island of 52.9 acres, rising only to 59 ft. The most abundant plant is Thrift, *Armeria maritima*, which forms a hummocky turf over most of the island (plate p 130). Formerly the tufts were waist high near Annet Head and, although some of the best areas have been reduced by fires, are still as much as a metre across. There are also large patches of Bracken, *Pteridium aquilinum*, about half a metre high on deeper soil on the sheltered side, with Lesser Celandine, *Ranunculus ficaria*, and Bluebell, *Endymion non-scriptus*, locally abundant. On two sandy areas, Sand Sedge, *Carex arenaria*, is dominant, while almost everywhere, and especially round the gull's nests, a grass, *Holcus lanatus*, is plentiful. Along the bouldery beaches the vegetation is mainly fleshy and Tree Mallow grows in rows like hedges. One small forest of this was found by Geoffrey Grigson to hide a prehistoric midden of mainly limpet shells. In May, when the Thrift flowers are freshly opened, Annet is a glorious colourful sight.

The island was frequented by prehistoric man who left barrows and puzzling ridge 'walls'. If it was ever cultivated it must have stood much higher in relation to the sea which now breaks right over it in storms. In the early nineteenth century it was used for grazing sheep (*Scillonian* **13,** 46) and cattle (Woodley, 1822 and North, 1850). Apart from a few fires, there is no evidence of human influence on the vegetation: the birds with their droppings and nest building have a much greater effect. Chickweed, *Stellaria media,* grows significantly near the landing place, where seed may have been introduced on a visitor's shoe; it is equally likely that it was brought by birds. Even the former cutting of Bracken for fuel by the inhabitants of St Agnes appears to have had no lasting effect.

THE ST HELEN'S GROUP

The St Helen's group with a total of 97 species, to the north of the archipelago, includes higher land permitting the growth of a some-what wider range of species. The islands have also suffered more human interference. Men-a-vaur is the most exposed but even this precipitous rock has *Cochlearia officinalis, Lavatera arborea* and *Atriplex* spp. Round Island, with its lighthouse high up above the sea, covers 9.5 acres. I have only toiled up the steps to the 136 ft top once, in 1938, and have no list of the flora but, as can be seen from the boat, Hottentot's Fig, *Carpobrotus edulis,* from South Africa is thoroughly established.

St Helen's, of 49.3 acres, rises to 144 ft with several acres of ground over 100 ft. In 1940 this was covered with Ling, *Calluna vulgaris,* and Fine-leaved Heath, *Erica cinerea,* which were destroyed by fires caused by incendiary bombs from enemy aircraft during the war. Further fires burnt throughout the dry weather of 1949 (*Scillonian* **24,** 128) and the vegetation has still not recovered. When I carried out a detailed survey of the island in 1957, the soil on the top was only one to two inches deep, grey, with granite chips, the most abundant plants were a plantain, *Plantago coronopus,* and stonecrop, *Sedum anglicum,* and most of the associated species were not ones likely to withstand much salt. Large areas of the sloping sides of St Helen's, the rocks, cliffs and flats are covered by the alien *Carpobrotus edulis,* which is spread here by gulls taking bits for their nests. The lower ground along the sheltered south-east side of the island has deeper soil, black with considerable raw humus and white grains of sand. Here Bracken starts at about half a metre high and gets taller beyond the ruined church to form a tangle interlaced with Bramble, *Rubus ulmifolius,* and with a varied flora. Salt-spray has much less influence here than on the islands discussed earlier.

There is another reason for the comparative richness of the flora—

St Helen's has a long history of human interference which has opened up the habitats to allow the establishment of additional species. This probably dates back to the Celtic monastery and the monks are likely to have grown grain in the small fields of which the walls remain. About 1669 there was one family living on the island (Magalotti, 1821). Borlase (1756) reported that the island was deserted, but the Pest House, for the landing of people with infectious diseases, was established by Act of Parliament of that year (plate p 65). In 1808–9 the commander of the guardship kept in the pool during the Napoleonic Wars kept a garden on the island (Woodley, 1822). Goats and deer were there in 1850 (North) and deer in 1865 (L'Estrange), while sheep ran wild in 1870. There are a few plants like Hemlock, *Conium maculatum*, and High Taper, *Verbascum thapsus*, which may have been introduced by man, but the main effect of his activities has been to keep back the development of natural communities on the higher ground, and to build up a deeper soil on lower, more sheltered areas.

Northwethel is a low island rising to only 50 ft, of 12 acres, and between St Helen's and Tresco. Most of it is covered with dense Bracken nearly a metre tall hiding a tangle of Bramble and very difficult to penetrate, and yet in one afternoon in May 1957 I listed 54 species. This is one more than the total for Annet which is more than four times the size. Whereas Annet is exposed to the south-west, Northwethel is sheltered by the high ground of Tresco so that the vegetation is taller, more luxuriant, and more varied. It was formerly used for grazing and this may have helped to improve the soil.

EASTERN ISLES

The Eastern Isles come next in order of groups with increasing diversity—111 species are recorded. This greater richness owes nothing to size as the largest, Great Ganilly, is only 32.6 acres, or to height, which is seldom over 50 ft and only attains 110 ft. The diversity arises from the position in the north-east of the archipelago sheltered from the west and south-west gales. The Eastern Isles form an interesting series with the number of species increasing with size (see Table 10) though with surprising anomalies in the plants represented. Thus, while the lists for Little Ganinick and Great Innisvouls are very much what one would expect from small islands, Great Ganinick with only 4.5 acres has a strange

Table 10   FLORA OF THE EASTERN ISLES

	No. of Species*	Acreage
Gt. Ganilly	74	32.6
Arthur	59	19.0
Nor-nour	39	3.3
Little Ganilly	37	6.0
Gt. Ganinick	35	4.5
Gt. Innisvouls	21	3.8
Little Ganinick	16	2.2

*These figures are based on a survey carried out by J. D. Grose, Mr. & Mrs. J. E. Dallas, and myself in 1938 and 1939 when the islands received equal treatment. Later additions are not included as the figures would no longer be comparable.

association of species, some of them normally woodland plants. Apart from Tresco it is the only place where Oak, *Quercus robur*, has been recorded in Scilly. It is one of the few places for Butcher's Broom, *Ruscus aculeatus*. Wood Spurge, *Euphorbia amygdaloides* is also there, and the Smallreed, *Calamagrostis epigejos*. No doubt high atmospheric humidity enables them to grow, but I am unable to suggest any special conditions to explain why so many uncommon plants are associated here. Nor can I explain why Great Ganinick is the home of the only Fumitory, *Fumaria capreolata*, to be found away from the five inhabited islands.

Nor-nour also has a surprisingly high total of 39 species for 3.3 acres. It has more plants requiring a fine state of soil erosion than are found on other islands of about the same size. Great, Middle and Little Arthur, joined together by boulders and blown sand to form a magnificent crescent round Arthur Porth (plate p 65), offer more varied habitats and more species (the total to 1970 is 73). This arises mainly from the large areas of blown sand, with Marram-grass, *Ammophila arenaria*, sheltering a richer flora, and the greater number of non-maritime plants including even a Sallow, *Salix atrocinerea*. Great Ganilly with 32.6 acres is the largest of the Eastern Isles and also the highest. It rises to 110 ft towards the north end, so that there is more shelter and more ground out of reach of spray under normal conditions. It consists of two granite masses joined together by a sandy 'neck'—like a miniature Samson. Here the land plants are more in evidence with sheets of Fine-leaved Heath, *Erica cinerea*, Common Ragwort, *Senecio jacobaea*, Golden Rod, *Solidago virgaurea*, and others.

Apart from the summer residence of kelp burners, the Eastern Isles have not been inhabited in historic times, but there is plenty of evidence of earlier occupation. On the larger islands there are not only barrows and megalithic graves but also remains of walls, and the recent excavations on Nor-nour have revealed occupation as recent as Roman times. Menawethan and Hanjague excepted, the islands are joined to St Martin's and St Mary's by very shallow sea, as may be seen from the launches at low tide, and this was formerly land. It was not until about 2,000 years ago that it was extensively submerged (see p 15) and the plants now growing on the Eastern Isles are mainly those which reached them across land, some of it probably cultivated, and have been able to persist on the higher granite hillocks which remained above sea level. Although Man has done little to disturb the islands in recent times, the present diversity of plants may owe something to earlier activities on Nor-nour and elsewhere.

TEAN

On Tean human influence on the vegetation is considerable, recent, and evident. The area is 39.5 acres, to which Old Man with 2.9 and Pednbrose with 1.4 must be added, and the number of species recorded is 119. In 1822 Woodley gave the acreage as seventy and land is still being lost to the sea. The island is made up of a series of granite mounds, of which the largest, Great Hill, rises to 100 ft, joined together by Head (rab) and Blown Sand. About twenty acres have soil deep enough for cultivation and much of this was walled off into fields of which some are shown on the Ordnance Survey maps.

The island was inhabited from very early times and remains of a long series of cultures have been found. The cottage which stands between East Porth and West Porth is on a site which was used intermittently for some nine-hundred years (Grigson, 1948), and especially by the Nance family who settled there in 1684 and were still cultivating and pasturing the island from St Martin's in 1822 and later. Borlase in 1756 reported fields of corn. Tean was grazed by cattle until 1945 (*Scillonian* **19,** 98–110) when there was also a donkey, but visitors can no longer expect the experience of J. D. Grose who, having spent a day there alone listing the flowers in 1939, was surprised to meet a bull face to face when the boat called to take him off.

The areas formerly cultivated are now overgrown with Bracken-Bramble tangle but the deeper soils are indicated by the abundance of Hogweed, *Heracleum sphondylium*. The effect of grazing is now almost lost but thirty years ago it was very evident in the short vegetation behind East Porth and West Porth where the best pasture was found. The former agricultural use is also suggested by the list of pasture plants not found on other uninhabited islands and likely to be encouraged by farming operations—*Trifolium pratense*, *T. campestre*, *Centaurea nigra*, *Lolium perenne*, and *Trisetum flavescens*. The last is not found elsewhere in Scilly. *Rumex pulcher* and *Hyoscyamus niger* are plants of disturbed habitats, while Common Mallow, *Malva sylvestris*, grows by the cottage just as it does round many a mainland farmyard.

SAMSON

Samson with 95.4 acres is the largest of the uninhabited islands and, with 127 species, is also the one most influenced by human activities. It consists of two granite hills rising to 100 ft, joined by Blown Sand, and with areas of Head (rab) on their slopes. Over the highest ground Ling and Fine-leaved Heath are often dominant, on the slopes on deeper soil Bracken-Bramble tangle covers large areas, especially on South Hill. The tops of the two hills are subject to terrific winds and in places the vegetation is so dwarfed that no plant is more than one and a half inches tall.

There may have been an early monastery here and possibly some prehistoric cultivation but, in any case, Samson was occupied intermittently more recently. It was uninhabited in 1579 (Godolphin) and 1650 during the Civil War, but one family had settled there by 1669 (Magolotti), it had a population of twelve in 1715, there were two households in 1751 (Borlase), and six, with thirty persons, in 1794 (Troutbeck). About 1834 Samson was inhabited by five families who grew a little corn (*Scillonian* **27**, 263–265), whereas in 1822 Woodley had reported seven families and thirty-four persons and said that earlier there were fine meadows on clay on the 'neck' joining the two hills. By 1850 the population was down to three or four households (North 1850 p 10) who were evacuated in 1855 (*Scillonian* **31**, 111–115). All these people could not live off the land but supplemented this by fishing and pilotage. When this failed their life was miserable in the extreme, and Augustus Smith forced the evacuation.

The human pressure on the vegetation of the island, much of it shallow soil over granite, must have been very great and there is still evidence of this. The walls round the old fields are easy to see in the spring before the Bracken grows up to hide them, and some of the enclosures have deep soil and luxuriant vegetation indicating that they were arable. As Geoffrey Grigson was the first to point out, on the eastern slopes of South Hill Bluebells are practically dominant within the enclosures and tend to stop at the boundaries where the soil was not cultivated. By the ruined houses on the north slopes of South Hill there are bushes of Tamarisk, *Tamarix gallica*, and Elder, *Sambucus nigra*, which have been planted, the first perhaps for shelter, the second for superstitious reasons. Cow Parsley, *Anthriscus sylvestris*, is also there and was probably introduced. There are also Primroses, *Primula vulgaris*, which are perhaps not native anywhere in Scilly.

After the island was depopulated it was used as a deer park (Scott and Rivington, 1870) and as pasturage for sheep and cattle from Abbey Farm (Tonkin and Row, 1906). Today human pressure is very considerable as large numbers of tourists are landed there almost daily during the summer and as many as two-hundred children have picnicked there in a day for school treats. The accumulation of sand on the neck, which Woodley reports as commencing towards the end of the eighteenth century, has continued and this now has an interesting dune flora. Lack of fresh water was one of the factors which, in contrast to Tean, made life on Samson so hard but there is a spring near the ruins and another, Southward Well, above the shore on the south-east corner. Both have a Starwort, *Callitriche obtusangula*, and Water Blinks, *Montia fontana* subsp *amporitana*, and about the second, along the shore, grows Greater Skullcap, *Scutellaria galericulata*, in its only station in Scilly.

The uninhabited islands of Scilly thus form a most interesting series showing an increase in the number of species related to size and hence decreasing dominance of maritime conditions. In Tean and Samson the number of species is further increased by the opening up of habitats by human activity so that additional plants could obtain a footing. There is a sharp contrast with the five inhabited islands where even the smallest, Bryher, has more than twice as many species as Samson, the largest uninhabited. This is, of course, mainly due to area and cultivation, but the presence of alien plants introduced from abroad is an important factor. There are many of these on the main islands but only three on the uninhabited— Hottentot's Fig, *Carpobrotus edulis*, from South Africa, and the fence plant, *Pittosporum crassifolium*, from New Zealand, which are both carried from island to island by birds, and the Mediterranean Tamarisk, *Tamarix gallica*, which has been planted on Samson.

The sea birds have also had a considerable influence on the vegetation of the uninhabited islands. The burrows of Puffins and Manx Shearwaters form centres of erosion, while the treading and manuring of sea birds round their nests is detrimental to most plants. Not to all; a few species thrive and become exceptionally gross and succulent in the presence of excess guano. Enormous plants of Sea Beet, *Beta maritima*, Scentless Mayweed, *Tripleurospermum maritimum*, Sorrel, *Rumex acetosa*, and Orache, *Atriplex glabriuscula*, are often found by the nests. In general, vegetation benefits from small amounts of nitrogen-rich manure but is destroyed by an excess, and around the larger sea bird colonies there is a zone browned and almost bare where few plants grow.

On Annet the vicinity of gulls' nests, thick with feathers and droppings, has coarse clumps of the grass Yorkshire Fog, *Holcus lanatus*, which is dominant, with Bracken common and Bluebells frequent. On Little Arthur and elsewhere, Sea Beet and Scentless Mayweed are usually the most common plants round the nests. On Samson gulls nest in large numbers on the slope on the east side of South Hill. Here Bracken, Sorrel, and Ground Ivy, *Glechoma hederacea*, seem best able to tolerate the concentration of nitrogen. On the Pembrokeshire islands, which are near to the mainland, Gillham (1956) has shown that gulls transport viable seeds of cereals and weeds. There is as yet no direct evidence of this in Scilly but they certainly spread the Hottentot's Fig to St Helen's and perhaps Hangman's Island by carrying pieces of stem to make their crude nests. These took root on the tops of rocks and grew.

Chapter Five

PLANT HABITATS

There are no easy precedents to help us in understanding the vegetation of Scilly. There are of course resemblances to other windswept islands and stretches of coast on the west side of England, Wales and Ireland but few of these have been written up by ecologists. Those that have, such as Gillham's accounts of the Pembrokeshire islands, offer comparisons with only limited areas in Scilly. Most of the natural vegetation forms complex communities merging into one another, while the man-made bulbfields have a weed flora unique in the British Isles and with a seasonal rotation more like that of Mediterranean arable.

To describe this in detail, illustrated with the several hundred habitat studies which have been prepared, would require a larger

Table 11   PLANT HABITATS

A   THE SHORE
      1.   Cliffs and rocky shores
      2.   Sand and shingle shores

B   COAST AND 'DOWNS'
      3.   Dunes
      4.   Maritime grassland
      5.   Thrift-turf
      6.   Heath
      7.   Granite Carns
      8.   Bracken communities

C   BRACKISH AND FRESHWATER HABITATS
      9.   Maritime ponds
    10.   Large 'freshwater' ponds
    11.   Pools
    12.   Freshwater marshes—the 'Moors'
    13.   Streams

D   WOODLAND
    14.   Small areas sheltered by trees

E   CULTIVATED HABITATS
    15.   Bulbfields
    16.   Arable under other crops
    17.   Gardens
    18.   Stone walls—'Hedges'

62

book than this one. The account which follows is a generalised review with emphasis on a few of the communities of exceptional interest. The terminology used by ecologists for comparable communities elsewhere will generally be avoided in view of the difficulties of precise application. The habitats are summarised in Table 11.

THE SHORE

Over half the coast of Scilly consists of low cliffs of granite or of Head (rab) some 6 to 12 ft (2–4 metres) high, separated from the sea by a narrow foreshore of rock and tumbled boulders interspersed with small patches of sand and shingle. Only in a few places, such as Peninnis Head, St Martin's Head and the extreme north of Tresco, are there granite cliffs falling almost straight into the sea. For about a quarter of the coast the beach is shingle or sand, often flanked by dune.

On the boulder-strewn shores the most abundant species is *Beta maritima*, always fleshy and often very large. *Solanum dulcamara* and *Tripleurospermum maritimum* are also abundant and the following very common: *Cochlearia danica*, *Sagina maritima*, *Spergularia rupicola*, *Crithmum maritimum*, *Cirsium vulgare*, *Carduus tenuiflorus*, *Sonchus oleraceus*, *Armeria maritima*, *Atriplex* spp., *Rumex crispus* and *Festuca rubra*. More locally, *Cochlearia officinalis*, *Lavatera arborea* and *Rumex rupestris* occur. All these have to endure frequent soaking in salt spray, and so most have thick fleshy leaves.

On St Martin's and Tresco there are long stretches of sandy beach, and shorter lengths on St Mary's, Bryher, Samson, Tean and other islands. On these beaches the zone immediately above extreme high water mark usually produces *Cakile maritima*, *Honkenya peploides*, *Salsola kali*, and various species of *Atriplex*, of which the silvery leaved *A. laciniata* is the most characteristic. Most sandy beaches then slope up more or less gradually to a belt of *Carex arenaria*, *Ammophila arenaria* and *Agropyron junceiforme*, which fix the blown sand sufficiently for additional species such as *Glaucium flavum*, *Eryngium maritimum*, *Calystegia soldanella* and *Festuca rubra* to become established. A few beaches, like those at Porthellick Bar and Wingletang Bay, are of fine shingle with a flora similar to that of the sand. Where the shingle is coarse, as at Periglis Bay, St Agnes, there are plants characteristic of both sandy and boulder beaches.

Most of these shore plants have buoyant fruits and seeds which remain viable in sea water for long periods. I have found that 10 out

of 18 fruits of the dock, *Rumex rupestris*, germinated after floating
in sea water for 181 days (Lousley, 1944 p 570) which is long enough
to be carried by the sea from island to island. Similarly *Cakile
maritima*, *Crambe maritima*, *Honkenya peploides*, *Calystegia soldanella*
and *Euphorbia paralias* are plants thought to be dispersed in this way.
The seeds of *Polygonum raii*, *P. maritimum* and *Euphorbia peplis*
must also withstand long immersion in salt water. These plants
appear for a time at rare intervals on shingly beaches, and the seeds
must either be there already buried under the shingle and brought
up after great storms, or else they are brought afresh from elsewhere.
Seedlings from seeds arriving at high water mark obtain their food
from the decomposing seaweeds thrown up with the tide.

DUNES

In Scilly the flora of the beaches grades into that of the coast and
in turn into that of the 'Downs'—the heathy higher ground. There
is seldom an abrupt change, and the change is very gradual indeed
when it is to coastal dune. There are good examples of dune at
Appletree Banks on Tresco, at Great Bay and elsewhere on St
Martin's, and Bar Point on St Mary's. There are smaller dunes at
the south end of Bryher, extreme north of Samson, and elsewhere.
It is only at Appletree Banks that they rise to high sandy hillocks,
the rest are either slopes where the sand has blown up over the
granite, as at Great Bay, St Martin's, rising to The Plains, or almost
flat as on Bryher. In all cases the following usually occur wherever
the dune is fairly stabilised: *Cerastium diffusum* (often very small),
Sagina maritima, *Erodium cicutarium*, *E. maritimum*, *Trifolium
arvense*, *T. repens* (often as var. *townsendii*), *Lotus corniculatus*,
Sedum anglicum, *Eryngium maritimum*, *Achillea millefolium*, *Senecio
jacobaea*, *Leontodon taraxacoides*, *Taraxacum* spp., *Armeria maritima*, *Anagallis arvensis*, *Centaurium erythraea*, *Myosotis ramosissima*,
Plantago coronopus, *Euphorbia portlandica*, *Agrostis stolonifera*,
Aira praecox, and *Festuca rubra*.

Appletree Banks, extending north along Pentle Bay, is the largest
of the dune systems. This has been formed from sand blown inland
from the beach building up the land to a considerable depth with
granite rocks and hillocks rising above the sand (Plate p 97). Since
the 1830's Marram Grass, *Ammophila arenaria*, has been planted
along the seaward side to arrest the sand, which is now built up by
the grass into high protective ridges. On these *Ammophila* is dense

INHABITED ISLANDS: (*top*) Eastern Isles, Middle Arthur with Little Ganilly behind and Martin's in background; (*bottom*) St Helen's, the Pest House, with bracken community in foreground

and almost the only vegetation. *Rubus ulmifolius* and *Scrophularia scorodonia* get established at an early stage and *Euphorbia portlandica*, *Sonchus oleraceus* and *Senecio jacobaea* are among other early arrivals. Occasionally the outer dunes are temporarily breached by 'blow-outs' and the danger of these is likely to be increased by the public treading out paths to the beach.

On loose sand behind these ridges some smaller species are present. Thus on a flat area at the south end of Pentle Bay, Tresco, the following occurred (*HS 203, May 1959): *Sedum anglicum* abundant; *Erodium maritimum* frequent; *Scrophularia scorodonia*, *Anagallis arvensis*, *Myosotis arvensis*, *Lycopsis arvensis* and *Euphorbia portlandica*, occasional.

Sometimes Sand Sedge plays an important part in consolidation of the soil, and the transition to dune heath, as in an open community on the west side of Carn Near Road (HS 204, May 1959): *Carex arenaria* abundant; *Viola riviviana*, *Aira caryophyllea*, *Erodium cicutarium*, *Anagallis arvensis*, *Rumex acetosella*, and *Euphorbia portlandica*, frequent; *Calluna vulgaris* (invading), *Euphorbia paralias* and *Centaurium erythraea* occasional; and *Myosotis ramosissima*, rare.

A hollow on sand between the Penzance Road and the greenhouses is representative of the open type of dune heath (HS 200, May 1959): *Calluna vulgaris* close cropped by rabbits, dominant; *Luzula campestris*, abundant; *Ornithopus pinnatus*, locally frequent; *Rubus ulmifolius* with low prostrate stems, *Viola riviniana*, *Veronica chamaedrys*, *Lotus corniculatus* and *Trifolium suffocatum*, occasional; *Teucrium scorodonia*, and *Ornithopus perpusillus*, rare. Eventually this develops into dense *Callunetum*. In the more sheltered areas of the older dune, Bracken becomes dominant, and on the exposed granite there is vegetation characteristic of rocks (see below).

Appletree Banks show a series of transitions from mobile dune to fixed dune, from dry to wetter habitats, from loose sand to Heath and to Bracken dominated areas, and to granite rock. It is unfortunate for the botanist that the dunes have been used for planting out exotic plants from the Abbey Gardens, some of which, like *Agapanthus praecox*, are spreading. Bar Point, St Mary's is almost equally interesting (Plate p 97). The sand here has blown up from Crow Bar and is ultimately piled up to a height of about 75 ft (23 m)

---

*HS=Habitat Study. These numbers refer to a series of vegetation studies of selected areas made by the author.

over the granite slope to the south. From north to south the sequence of communities is:

1. *Zostera marina* submerged in shallow sea.
2. Pure community of *Agropyron junceiforme* just above HWM.
3. Scattered plants of *Ammophila arenaria* with *Calystegia soldanella* abundant, and other shore species.
4. *Ammophila* dominant with occasional plants of *Calystegia soldanella, Hypochoeris radicata, Jasione montana, Pteridium aquilinum* and *Anthoxanthum odoratum*.
5. Consolidated sand with broken sea shells by the cable sign. *Galium verum* abundant; *Carex arenaria, Plantago lanceolata, Hypochoeris radicata, Plantago coronopus*, frequent; *Bellis perennis* and *Erodium cicutarium*, occasional.
6. Open Bracken, 0.5 m tall, about 8 stems to the square metre on consolidated sand with considerable humus. Here there is a much longer list of associated species, of which only one, *Carex arenaria*, has maritime associations.
7. Gorse thicket, with *Ulex europaeus, Pteridium aquilinum* and *Rubus ulmifolius* all abundant, and *Rubia peregrina, Lonicera periclymenum*, and *Veronica chamaedrys*, frequent.

This account is based on a survey made in July 1954 (HS 69) and the sequence of communities can still be seen in spite of disturbance due to relaying of the cable, removal of sand, and dumping of rubbish.

As this example shows, where there is dune the change from beach vegetation is usually gradual, but most of the coast of Scilly is low cliff and here the change is abrupt. Along the edge of the cliff there is commonly a belt of grassland, usually mainly composed of fescues (*Festuca rubra* and *F. ovina*). With these are commonly associated *Armeria maritima, Plantago coronopus, Leontodon taraxacoides, Sedum anglicum, Daucus carota*, and *Achillea millefolium*. *Holcus lanatus* is also usually present and in the moister places, and near seabirds' nests, this is often dominant. Elsewhere *Armeria maritima* may be dominant, forming hummocks when alone, but a low springy turf when mixed with fescues and subject to trampling.

THE DOWNS

The Downs are one of the great attractions of Scilly. On this higher exposed land, where the soil is thin over granite, Ling, *Calluna vulgaris*, is dominant over large areas. This maritime or 'waved' heath is a western type developed in places subject to frequent gales and is very different from the usual *Callunetum* of the mainland. Less than 15 cm high, knotted and gnarled, windpruned and eroded, there are fine examples on Castle Down, Tresco; Shipman Head Down, Bryher; and Chapel Down, St Martin's. There are many acres of it on St Mary's and St Agnes (Plate p 130) and it occurs also on Samson, St Helen's and Great Ganilly.

An area on the top of Shipman Head Down is an example (HS 83, August 1956). Here *Calluna* is prostrate in ridges with heavy wind erosion so that the root systems are exposed on the western side, and the flowering shoots prostrate directed towards the east. Here on ground over 100 ft above sea level and exposed to the full force of the gales, all plant life must hug the ground. The only associated species are occasional plants of *Erica cinerea, Leontodon taraxacoides, Armeria maritima, Lotus corniculatus* and *Sedum anglicum*. Under such conditions growth must be slow but in time a peaty soil is built up. In the past this was cut by islanders for fuel and it burns freely in situ when fires are accidentally started after a spell of dry weather.

Erica cinerea is commonly associated with *Calluna*, but only occasionally becomes itself dominant, as, for example, on St Helen's and towards the south end of Castle Down. Other common associates include *Potentilla erecta, Galium saxatile, Pedicularis sylvatica,* and *Agrostis tenuis*.

Very locally, on flat ground with a thin soil over granite, small forms of *Erodium maritimum* or *Plantago coronopus* may be dominant. These are fascinating little plants, usually less than 3 cm across. The plantain in this state has leaves which are almost entire and very unlike the fleshy divided leaves of the same species growing on deeper soils. The soil contains granite shingle in varying proportions and is acid (*pH* about 4.8), and water stands in the winter. Both species sometimes occur together as, for example, on open ground facing west on the ridge just by the end ruined cottage on South Hill, Samson. The soil here is up to 8 cm deep over granite and in this wind-swept spot no plants are more than 4 cm tall. The open vegetation consisted of (HS 12, June 1954): *Erodium maritimum* and *Plantago coronopus* co-dominant as small plants about 2 cm in diameter, *Rumex acetosella* abundant, *Sagina procumbens* frequent, and *Aira praecox* and *Poa* sp. occasional—the only species in the square metre examined.

The vegetation of the Carns (the granite outcrops which stand up sharply on the downs and headlands) also includes a number of heathland species. *Calluna* is usually present and often *Umbilicus rupestris*. Carn Grigland, on St Agnes, may be given as an example although these two species were not observed when it was listed in August 1956 (HS 118). On the shallow soil between the rocks there were a few small scattered plants of *Ulex europaeus*, and *Plantago coronopus* was abundant. *Leontodon taraxacoides*, *Spiranthes autumnalis*, *Sedum anglicum*, *Holcus lanatus*, *Rumex acetosella*, and *Lotus corniculatus* frequent, *Erica cinerea*, *Ornithopus pinnatus*, *O. perpusillus*, and *Plantago lanceolata*, occasional, and there were a few plants of *Anagallis arvensis*.

BRACKEN COMMUNITIES

Bracken is dominant locally on all the islands except the smallest. Whereas *Calluna* can thrive in the most exposed situations over very shallow soils, *Pteridium* needs some shelter from the worst of the winds and salt spray, and a deeper soil. The need for shelter and deeper soil can be seen on coasts where depressions alternate with ridges, as, for example on Tresco between the quay and Cromwell's Castle. Here *Calluna* is on the ridges and the depressions are green with Bracken. That is so in a normal summer, but in August 1970

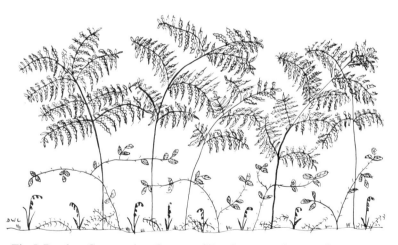

Fig 5 Bracken Community: Canopy of Bracken, *Pteridium aquilinum*, providing shade during summer; understorey of Bramble, *Rubus ulmifolius;* herb-layer of Bluebell, *Endymion non-scriptus,* and other mainly spring-flowering species

two successive gales whipping the salt spray on to the coast, turned the Bracken brown here as elsewhere in Scilly. On more exposed islands, such as Annet, Bracken suffers so much from spray every year that it is surprising that it can persist even in its rather dwarfed state.

Bracken in Scilly does not attain its maximum development until July and it can then reach a height of over 2 m in a few sheltered spots—such as a few places on the east coast of St Mary's. It is then that visitors become painfully aware that there is an understorey of *Rubus ulmifolius* invisible below the fronds and ready to gash their flesh and tear their clothes. In spring, before the new fronds start to grow, the ground is blue with the flowers of Bluebells, *Endymion non-scriptus*.

This *Pteridium-Rubus-Endymion* association consists of three layers (Fig 5). The upper one of Bracken fronds is present only during the summer and it then acts like the canopy of trees in a wood, cutting off light from every plant below, and maintaining a relatively still and humid atmosphere. It is only occasionally that other plants rise above the Bracken. Its rhizomes run deep in the soil. The understorey, of *Rubus ulmifolius* and a few other species, has only reduced light during the summer. These plants take advantage of small gaps in the canopy and here and there rise above it. The ground layer has full light early in the year and the advantage of humus provided by the decaying fronds. In the vernal aspect Bluebells are dominant and some of the species present are woodland on the mainland.

The tangle on the south-east side of St Helen's near the Pest House (Plate p 65) is an example floristically richer than most (HS 127, May 1957). This is on Head and sand, with a black soil with considerable raw humus (*pH* 6.7). The following species were noted:

CANOPY

Pteridium aquilinum dominant, 0.5–1 m tall. One clump of *Ulex europaeus*.

UNDERSTOREY

Rubus ulmifolius dominant with the following growing up to this level: *Euphorbia amygdaloides* locally abundant, *Scrophularia scorodonia* and *Heracleum sphondylium*, common; *Digitalis purpurea* and *Silene dioica* frequent; *Lonicera periclymenum* occasional; *Rumex crispus* and *Rubia peregrina* rare.

HERB LAYER

Endymion non-scriptus abundant and dominant; *Glechoma hederacea, Ranunculus ficaria, Galium aparine, Carex arenaria*, and *Brachypodium sylvaticum* abundant; *Lotus corniculatus, Rumex acetosa, Poa subcoerulea, Vicia angustifolia* and *Ranunculus repens*, common; *Holcus lanatus, Veronica chamaedrys, Myosotis ramosissima*, frequent; *Arrhenatherum elatius, Agrostis stolonifera*, and *Umbilicus rupestris*, occasional; *Hypochoeris radicata*, and *Arum neglectum*, rare.

Pteridium-Rubus-Endymion occurs on all the islands except the very small ones. On some of the uninhabited, like Northwethel, it is extremely difficult to penetrate; where there are more visitors, as between Watermill Cove and Blockhouse Point, St Mary's it is traversed by narrow tracks. It grows equally well on Head or sand provided there is shelter and fairly deep soil. Occasionally it merges into gorse thickets of *Ulex europaeus* and *U. gallii*. These species are sometimes dominant over small areas.

BRACKISH AND FRESHWATER HABITATS

There are a few pools on the coast which are brackish, especially after exceptional storms which cause the sea to break in. Great Pool and Little Pool on the west coast of Bryher have a rather interesting contrast. The former is almost choked with *Ruppia maritima* and *Glaux maritima* while *Juncus gerardii* is abundant round the edge. These are all plants which can thrive under salty conditions, as also can *Spergularia marina* and *Chenopodium rubrum* growing in cut-outs in the turf by the pool. In nearby Little Pool, *Apium inundatum* and *Ranunculus baudotii* are abundant in the water with a *Callitriche*, and *Ranunculus flammula* and *Hydrocotyle vulgaris* round the edge— very much less evidence of salt. Pools on Pool Green, St Martin's are similar, having *Ranunculus baudotii* and *Callitriche intermedia*. At Big Pool, St Agnes, *Scirpus maritimus* is abundant and *Juncus gerardii* the only other species listed which suggests brackish influence.

Abbey Pool, on the sand of Appletree Banks Tresco, has a far richer flora, with no species which are markedly halophytic. There is a broad margin of black fine mud enriched by the droppings of ornamental waterfowl (*pH* 5.1) and the following plants occur (HS 30, June 1954; HS 109, August 1956; revised September 1970): On the exposed mud *Hydrocotyle vulgaris, Anagallis tenella, Radiola linoides, Littorella uniflora, Potentilla anserina*, a small form of

Ranunculus flammula, and *Elatine hexandra* are all abundant, *Anagallis arvensis*, *Centaurium erythraea*, *Lotus uliginosus* and *Peplis portula* are common; *Anagallis minima*, *Baldellia ranunculoides*, *Eleocharis uniglumis*, *Galium palustre*, *Gnaphalium uliginosum*, *Juncus bufonius*, *J. bulbosus*, *J. articulatus*, *Myosotis secunda*, *Sagina procumbens* and *Scirpus cernuus* are frequent; *Callitriche stagnalis*, *Chenopodium rubrum*, *Poa annua* and *Samolus valerandi* are occasional and *Polygonum hydropiper* rare. This margin is maintained and enriched by the grazing of the ornamental birds but they allow few plants to grow in the water where rare charophytes are sometimes found.

Great Pool, Tresco, has a less exciting flora. The pool is over half a mile long, protected from the sea by a sandbank at Abbey Farm at the north end, and by wider dunes in the south. The water is fresh, but the marginal belt of *Phragmites australis* is increasing and there is now little open water. In this *Ranunculus baudotii*, *Myriophyllum alterniflorum* and *Potamogeton pectinatus* have been found. On the margins the reed is building up the silt so that the water becomes shallower and carr develops. There is a good example of this on the promontory in the south-west corner, where there is spongy water-logged soil rich in humus with *Osmunda regalis*, *Oenanthe fistulosa*, and *Juncus effusus* growing between the rather widely spaced *Phragmites* stems. The next stage in the succession is the withy bed with sallows and a springy soil, waterlogged in winter and firm with a strong odour of decaying vegetation in summer. Here there is *Iris pseudacorus*, *Hydrocotyle vulgaris*, and *Silene dioica* and an under-growth of brambles, *Rubus* spp. merging into drier ground with *Ranunculus repens*, *Rumex sanguineus*, *Iris foetidissima*, *Ranunculus ficaria* and *Scrophularia scorodonia*. *Typha latifolia* is abundant at the south end of the pool, and *Osmunda regalis* extends on to the dunes. Great Pool is screened from Pool Road by *Salix cinerea*, planted trees and *Phormium tenax*.

On St Mary's there is a large sheet of open water on Higher Moors at Porthellick. This is separated from the sea by a shingle bar. At the seaward end of the pool there are a few salt-enduring species such as *Juncus maritimus*, *Scirpus maritimus*, *Glaux maritima*, *Samolus valerandi* and *Ranunculus baudotii*, and *Phragmites australis* which extends all the way round in an almost pure community. Sea water sometimes enters Porthellick Pool through the artificial drain at high tide, and may even introduce marine fish, but it has little effect on the vegetation, except that salt-enduring species are more plentiful near the drain. *Ranunculus flammula, Oenanthe crocata, Osmunda regalis, Iris pseudacorus, Mentha aquatica, Lycopus europaeus* and a reedy form of *Agrostis stolonifera* are among the marsh species found here. *Typha latifolia* also occurs.

There are a few smaller pools on St Mary's. One in a field below Carn Gwaval in Old Town Marshes is interesting because Curnow found *Ruppia maritima* and *Potamogeton pectinatus* there in 1876. These no longer occur but *Callitriche obtusangula* and *Ranunculus baudotii* are abundant. Two contiguous very shallow ponds at the top of Watermill Lane, near Lower Newford, are over 100 ft above sea level and over one-third of a mile from the nearest sea. They are the best examples of small non-brackish pools and the plants present include (HS 112, August 1956): *Lemna minor, Apium inundatum, Ranunculus baudotii, Nasturtium officinale* and *Glyceria declinata* in the water, and *Stellaria uliginosa, Juncus bulbosus, J. bufonius, J. acutiflorus, Polygonum hydropiper, P. persicaria, Eleocharis palustris, Rumex conglomeratus* and *Myosotis repens* in wet ground.

MARSHES

The two large areas of marshland are known as 'moors', and are on St Mary's. Higher Moors extend from Porthellick Pool almost to Holy Vale; Lower Moors from Old Town Bay, past Rose Hill to Porthloo Pool. Both are on low ground rising little above sea level, on alluvium, wet, rushy and often overgrown, the more so since their use for grazing has virtually ceased. Now their main use is for pumping stations to provide extra water for the increasing number of visitors. They are divided into rough fields separated by drainage ditches, hedges and a few clumps of sallow. In the wetter parts *Phragmites australis* and *Iris pseudacorus* are abundant and sometimes locally dominant.

No attempt will be made to describe all the communities present but several of them are of special interest. On entering Lower Moors by the short lane from Old Town, the first field to the left has the endemic *Juncus maritimus* var. *atlanticus* as the dominant plant. Associated species include (HS 5 June 1954): *Hydrocotyle vulgaris* abundant; *Galium palustre* and *Oenanthe fistulosa* common; *Lychnis flos-cuculi, Ranunculus flammula* frequent; *Juncus articulatus* and *Holcus lanatus* occasional; *Galium uliginosum, Prunella vulgaris* and

Cirsium palustre, rare. In part of this field the soil is acid, and the following species were recorded in a square metre by the track, where the *pH* of the soil was 5.1 (HS 6, June 1954): *Eleocharis multicaulis*, *Sieglingia decumbens* and *Carex nigra*, abundant; *Hydrocotyle vulgaris*, *Ranunculus flammula* and *Anthoxanthum odoratum*, common; *Agrostis canina* occasional; *Carex demissa* and *Cynosurus cristatus*, rare.

On Higher Moors there is a remarkable tussock development of the sedge *Carex paniculata* which in August 1906 attracted the attention of W. W. Smith (later Sir William Wright Smith) of Edinburgh (Smith, 1907) This is on the south side of the road between Tremelethen and Carn Friars and is hidden in the *Phragmites* swamp. These enormous tree-fern-like tussocks appear to form only when permeated from an early stage with the stems of the grass, *Phragmites australis*, and/or Bracken, *Pteridium aquilinum* (see p 285). On its own Smith claimed that the sedge formed tussocks less than a foot high. Today they are much taller, but the compound tussocks no longer attain 10 ft.

Immediately opposite, on the north side of the road, there is a small area of a few square metres where until recently the white heads of Bog Cotton, *Eriophorum angustifolium* drew attention to exceptionally acid conditions. Other plants characteristic of bogs included *Hypericum elodes*, *Potamogeton polygonifolius*, *Carex echinata*, *Epilobium palustre*, *Stellaria uliginosa* and *Anagallis tenella*. Some of these are still there but are very scarce and becoming crowded out by coarse vegetation. There was no Sphagnum Moss but the black soil was so acid (*pH* 5.5) that the flora was almost that of a bog. On both Lower and Higher Moors *Osmunda regalis* is abundant and very fine.

There is only one stream worthy of the name in the whole of Scilly. This rises a little south of Holy Vale, flows through Higher Moors and enters the sea at Porthellick. In half a mile it falls less than 25 ft and is so sluggish that it has cut a bed barely below the level of the surrounding marsh. The flora is that of the wetter marshes: *Nasturtium officinale*, *Epilobium palustre*, *Oenothera crocata*, *Angelica officinalis*, *Lythrum salicaria*, *Myosotis secunda* and *Iris pseudacorus* are the most conspicuous species present. The stream would become completely choked with the silt it brings down were it not cleaned out regularly. There is a drainage ditch across Lower Moors which is hardly a stream, and a few wet roadside gutters where interesting plants like *Sibthorpia europaea* and *Wahlenbergia hederacea* were found before the roads were surfaced with asphalt.

'WOODLAND'

There is no natural woodland in Scilly today. There is ample evidence that Oak, *Quercus robur*, was formerly plentiful (see p 205), that Hazel, *Corylus avellana* occurred, and probably also Ash, *Fraxinus excelsior*. These, together with perhaps other trees and shrubs, were destroyed long ago to meet the demand for timber and fuel so much needed by the islanders and crews of visiting ships. There are, however, a few stunted gnarled trees of *Q. robur* in Abbey Gardens, which may possibly be on the site of an old Abbey Wood of which there is a tradition. This is supported by the presence of a few woodland herbaceous species such as *Cardamine flexuosa*, *Carex divulsa*, *C. sylvatica* and *Veronica montana*, which do not occur elsewhere in the islands. The absence of old woodland explains why Scilly is so deficient in woodland herbs apart from the few which thrive under the Bracken canopy.

In recent years, however, there has been rapid development in elm copses. The best of these is above Old Town Church, St Mary's, which I was shown by P. Z. MacKenzie (HS 306, May 1967). Here there is dense growth of *Ulmus procera* from suckers over terraced ground believed to be under cultivation until thirty to forty years ago. *Heracleum sphondylium*, *Endymion non-scriptus*, *Allium triquetrum* and *Hedera helix* are common; *Endymion hispanicus*, *Anthriscus sylvestris*, *Dryopteris filix-mas*, *Humulus lupulus*, and *Iris foetidissima*, local; *Urtica dioica*, *Silene dioica*, and *Pteridium aquilinum* occasional, while *Senecio mikanioides* has spread from the churchyard to the edge. This is the nearest to a woodland flora which Scilly can produce.

In Lower Moors near Low Pool there is an example of an elm copse of a wetter type. Here *Ulmus procera* is again dominant and spreading from suckers, accompanied by the following (HS 307, May 1967): *Athyrium filix-foemina* common and very fine, and occasional *Rubus* sp., *Cardamine pratensis*, *Hedera helix*, *Dryopteris filix-mas* and *D. dilatata*. In a dry-type elm copse on the Garrison (HS 310, May 1967) *Ulmus procera* was associated with:—*Endymion non-scriptus* and *Allium triquetrum* abundant; *Hedera helix*, *Umbilicus rupestris* and *Smyrnium olusatrum* common; *Silene dioica* frequent; *Brachypodium sylvaticum*, *Galium aparine*, and *Ranunculus ficaria* occasional; *Rubia peregrina* and *Rubus* sp. also present.

These and other elm copses on St Mary's have developed quickly and recently.

Another new habitat type is that of the coniferous shelter belts. A belt of old trees of *Pinus radiata* near Higher Trenoweth, in the north of St Mary's, is representative. Very little grows on the litter of needles but occasional *Umbilicus rupestris* and *Polypodium vulgare*. and more rarely *Rubus* sp. and *Silene dioica*.

Sallow copses have no doubt been in Scilly since historic times. There are several in Higher and Lower Moors, St Mary's, and about Great Pool, Tresco. One on Lower Moors, near Rose Hill, on the site of Low Pool, which until recently was open water, is representative. Here bushes of *Salix cinerea* about 2 m tall are dominant (HS 76, July 1954) with *Iris pseudacorus*, *Ranunculus repens* and *Poa trivialis* abundant; *Hedera helix*, *Galium aparine*, *Lonicera periclymenum* and *Dryopteris dilatata* frequent. Other Salices are sometimes involved and water levels vary widely during the year and from copse to copse.

CULTIVATED HABITATS

Over 600 acres are devoted to growing bulbs for the cut-flower industry or for sale as bulbs and, as explained earlier (see p 37) there has developed a very special type of weed flora characterised by species which grow through the winter. Some of these are unknown on the mainland, others occur here in far greater abundance. Although these bulbfield communities have a general similarity they vary from field to field and island to island. This is best illustrated by a few examples.

A bulbfield by the lane leading to Old Town Church, St Mary's, (HS 171, May 1957) on Head had the following: *Oxalis pes-caprae*, *Briza minor*, *Aira caryophyllea* and *Juncus bufonius* abundant; *Ranunculus muricatus*, *Bromus diandrus*, *Bromus willdenowii*, *Veronica arvensis*, *Trifolium dubium*, *Sonchus oleraceus*, *Silene gallica*, *Valerianella locusta* and *Poa trivialis*, common; *Medicago arabica*, *M. polymorpha*, *Cerastium glomeratum*, and *Fumaria boraei* occasional; *Papaver dubium* rare. This may be contrasted with another bulbfield on Head near Old Town at Porth Minick. Here (HS 300, May 1967) a small bulb-square was surrounded by a fence of *Pittosporum* about 5 m tall and *Chrysanthemum segetum* was dominant over about half the field away from the shaded edges where *Allium triquetrum* replaced it. *Oxalis pes-caprae* and *Bromus diandrus* were abundant;

Ranunculus muricatus, Cerastium glomeratum, common; *Poa annua* (a large form), *Sonchus asper, Medicago polymorpha, Erodium moschatum,* and *Ranunculus parviflorus* frequent; *Medicago arabica, Fumaria boraei, Ranunculus repens, Valerianella locusta, Galium aparine, Veronica persica, Allium babingtonii, Gladiolus byzantinus,* and *Heracleum sphondylium,* occasional; *Crepis vesicaria, Rumex obtusifolius* and *Senecio vulgaris* rare. Thus although the two most conspicuous species in this field were absent from the first, the associated plants included a number which were the same.

Bulbfields on sandy soils vary greatly. Sometimes they are taken into cultivation for only a short time and then abandoned and in such cases, as with one at Pelistry Bay (HS 79) all the weeds may be native species—there is insufficient time for the aliens to spread. As a more typical example a bulbfield on sand below Bant's Carn Battery, St Mary's, may be cited (HS 304, May 1967). Here *Gladiolus byzantinus, Montia perfoliata, Endymion hispanicus,* and *Allium triquetrum* are the most abundant species; *Valerianella locusta, Ranunculus parviflorus,* and *Arabidopsis thaliana* common; *Callitriche stagnalis* (on 'dry' ground) and *Crassula decumbens* locally common; *Erodium moschatum, Medicago polymorpha, M. arabica, Allium babingtonii, Aphanes microcarpa,* and *Oxalis pes-caprae* occasional. On St Martin's there are many bulbfields on sandy soils and these usually have a higher proportion of native arenicolous species. Thus a field below the old school (HS 41, June 1954) had *Silene gallica, Polycarpon tetraphyllum, Vulpia myuros, Bromus diandrus, Briza minor, Geranium molle* and *Aira caryophyllea* abundant; *Trifolium dubium, Erodium cicutarium, Lycopsis arvensis, Anagallis arvensis* and *Viola arvensis* common; and *Rumex acetosella* occasional. A bulbfield adjoining the church had an unusual weed population (HS 311, May 1967). The most abundant species was *Ranunculus marginatus* which coloured the whole field yellow, and *Montia perfoliata* was almost equally plentiful. *Ornithogalum umbellatum* and *Cerastium glomeratum* were common; *Veronica persica* and *Anthriscus caucalis* frequent; *Gladiolus byzantinus, Plantago lanceolata,* and *Senecio vulgaris* occasional, and there were a few plants of *Sisymbrium officinale.*

On Tresco most of the bulbfields resemble fairly closely one or other of the examples already given, but in the fields adjoining Pool Road, *Nicandra physaloides, Portulaca oleracea* and *Amaranthus retroflexus* are often abundant. Here too, as in many fields on

this and other islands, *Oxalis articulata* often occurs in great quantity
On St Agnes and Bryher there is rather less variety in the bulbfield
weeds but they follow the same general pattern.

Other tillage crops occupy only about a third of the acreage
devoted to bulbs and are usually grown to 'rest and clean' the soil
for a further bulb crop. Early and maincrop potatoes, and roots are
mainly used for this purpose and cultivated with a view to reducing
the heavy population of weeds. This is seldom entirely successful
but tends to produce a temporary abundance of coarse species with
which we are familiar on the mainland. Thus a field of beet near
Old Quay, St Martin's, on sandy soil (HS 40, June 1954) had
Chenopodium album, *C. murale* and *Urtica urens* abundant; *Atriplex
patula* and *Fumaria boraei* frequent; and a few plants of *Medicago
arabica* and *Lamium amplexicaule*. These are less successful when the
field returns to the less disturbed conditions of a bulbfield, and more
interesting weeds crowd them out in the seedling stage.

Clover and grass leys are usually fairly free of weeds but grasses
such as Timothy, *Phleum pratense*, and Meadow Foxtail, *Alopecurus
pratensis*, common on the mainland, seem to occur in Scilly only
where introduced in seed mixtures. The weeds of gardens include
many of those of bulbfields and also native species grown exceptionally large. *Lotus hispidus*, *Ornithopus pinnatus* and *O. perpusillus* are
examples. Abbey Gardens, Tresco, as might be expected, have
weeds of their own from plants which have been introduced for
cultivation.

Stone walls are to be seen along many of the roads and also round
some of the larger fields. They are built of squared granite blocks
joined together with earth and some three to five feet thick, the
thicker ones often filled with soil. In all cases the tops become
covered with earth, and many of the walls appear to be of great age.
Here they are known as 'hedges' and they resemble those of Cornwall
except that the vegetation is usually less rank.

On the sides and tops of the Scillonian 'hedges', Fumarias, *Lotus
corniculatus*, *Umbilicus rupestris*, *Sedum anglicum*, *Hypochoeris
radicata* and *Allium triquetrum* make a colourful display. Other
wall plants include *Arenaria serpyllifolia*, *Polycarpon tetraphyllum*,
Hypericum humifusum, *Jasione montana*, *Rumex acetosella*, *Polypodium vulgare* and *Asplenium billotii*. Ralfs, Beeby and other early
botanists found many interesting plants such as *Trifolium suffocatum*
and *T. glomeratum* on the tops of these stone-walls which I have

failed to find there. In their time they must have resembled in their flora the famous stone-walls used as footpaths in the Lizard-Kynance area. It is possible that in Scilly the older broader walls have in some cases been destroyed on account of the valuable ground they occupied and harbouring weeds. One of the reasons weeds reproducing from bulbs, such as *Allium triquetrum* and *Oxalis pes-caprae*, have increased so rapidly in spite of the efforts of the farmer is that they have found sanctuary in the walls while the fields were being cleaned.

Chapter Six

DISCOVERY OF THE FLORA

The dangers and discomforts of the sea crossing have been sufficient to discourage all but the most determined of the early visitors to Scilly. This explains the late discovery of the interesting flora and the few visits from botanists until after the last war. Modern boats provide a standard of comfort and reliability which gives little idea of the conditions which prevailed until recently and since 1937 an air service has provided an alternative for those unable to face the sea. Large numbers of visitors now cross every year, many of them botanists.

When Turner and Dillwyn published their *Botanist's Guide* in 1805 they were unable to include a single record from the Isles of Scilly, although many of the Cornish rarities had been known for a century or more. The only pre-twentieth century records were a few incidental mentions of plants in general works, such as Robert Heath, 1750 and Borlase, 1756, and a record of a plantain in Gerarde's *Herbal*, 1633.

In 1813 the islands were visited by William Jackson Hooker (1785–1865), who was to become Sir William Hooker, the famous first Director of the Royal Botanic Gardens, Kew. He crossed from Cape Cornwall in a Trinity House boat on 26 April, and returned on 5 May after visiting all the larger islands. At this time of the year many interesting flowering plants could have been seen but all he reported were some 'wretched plants of elm and *Tamarix gallica*'. Mosses and lichens he found 'parched almost to a cinder', and the only records of the visit are a graphic account of the state of the islands (Dawson Turner Correspondence, Kew), a sketch of the ruins of Tresco Abbey and a few specimens including Primrose Peerless, *Narcissus x medioluteus*, which is important as the earliest reliable record of a Narcissus in Scilly. It was a great opportunity lost.

The first botanical list appeared in 1821 as a supplement to a small book on the climate of Penzance (Bree, in Forbes, 1821). This gives 14 rather rare species and was prepared by the Rev. William Thomas Bree (1787–1863), a Warwickshire botanist, who spent some months in west Cornwall. His records include a Figwort, *Scrophularia scorodonia*, Elecampane, *Inula helenium*, and Ivy-leaved Bellflower, *Wahlenbergia hederacea*, and he gave localities. Five years later the islands were visited by Francis King Eagle (1788?–1856) whose specimen of White Mignonette, *Reseda alba*, dated 1826, is at the Natural History Museum. He was mainly a bryologist and I have been unable to trace any records of other flowering plants he may have found on this visit. In 1838 a Miss Matilda White discovered Jointed Birdsfoot, *Ornithopus pinnatus*, on Tresco, and new to England. Her letter to Sir William Hooker also records Evergreen Alkanet, *Pentaglottis sempervirens*, from Tresco.

In September 1848 the islands were visited by Edward William Cooke, RA (1811–80), a marine painter, who cultivated ferns at his residence 'The Ferns', Kensington. He produced a most competent account of those he found in Scilly, covering 15 species, with localities for most (North, 1850). In 1845 a few plants were collected by 'A. Hamburgh', who was almost certainly Albert John Hambrough (1820?–61).

In 1852 no fewer than four good botanists visited the archipelago, and between them they put Scilly on the botanical map. In August of that year, Joseph Woods (1776–1864), one of the most travelled botanists of his day and author of the *Tourist's Flora*, found Sea Knotweed, *Polygonum maritimum*, Purple Spurge, *Euphorbia peplis*, and Eel-grass, *Zostera marina*. In 1852 John Ralfs (1807–90) commenced the series of visits which culminated in the manuscript floras now at Penzance (Marquand, 1893), and these are discussed below. By far the greatest contribution came from the Misses Millett who spent five weeks of June and July listing the flora and preparing models of the prehistoric monuments.

Miss Louisa Millett (1801–71) and Miss Matilda Millett (1805–55) listed 144 flowering plants and 16 ferns, of which 121 phanerogams and 3 ferns are accepted as first records. The accuracy of their list deserves the highest praise and there are probably very few errors which the text-books then current would have enabled them to avoid. It is unfortunate that they included no localities though it is stated that most of the botanising was done on St Mary's. Their account appeared in the *Report of the Penzance Natural History and Antiquarian Society* for 1852, but is dated 21 October 1853, and may not have been published until 1854 or even later. I have cited the records as 'Millett, 1852'.

In about 1857 Sir Joseph Dalton Hooker (1817–1911), in the company of Professor William Henry Harvey (1811–66) and the great nurseryman James Veitch (1815–69) visited Tresco to see the Cape and Australian plants which Augustus Smith was growing (*Botanical Magazine, tab.* 5973). I have so far failed to trace any records of specimens from this visit, but *Escallonia macrantha* and other garden plants reached Tresco from Veitch.

On 21 June 1862 Frederick Townsend (1822–1905) crossed to Scilly and put up at Tregarthen's, the hotel run by the captain of the boat. It was 'a somewhat boisterous passage of 5 hours', which may explain the note in his manuscript that he was not well during his stay, which ended on 1 July. During these ten days Townsend compiled the remarkable annotated list of 348 species, which was published in the *Journal of Botany* for April 1964. Doubtless he had some preliminary help from A. J. Hambrough and Joseph Woods, and being 'a connection' of the Lord Proprietor, Augustus Smith, must have smoothed his path in the islands, but the volume of work achieved in such a short time was highly creditable. The list includes 190 first records. Townsend's account included rushes, sedges and grasses which had been practically ignored by earlier writers and it was highly 'critical' throughout. Unfortunately the shortness of the visit must have led to haste and carelessness, and the frequencies appear to be mere guesses from memory and are often misleading. There are 21 almost certain errors, many of which were detected by Townsend himself after publication and altered in his own copy of his paper.

Townsend's paper stimulated further interest in the islands. In 1869 they were visited by Marmaduke Alexander Lawson (1840–96) who published a paper the following year claiming twenty-three additions and two corrections to Townsend's list. Of these only seventeen were really new, and there were further probable errors. In July 1873, John Ralfs came again, accompanied by Richard Vercoe Tellam (1826–1908) and William Curnow* (1809?–87). They discovered Cretan Mallow, *Lavatera cretica.* In the same month William Hadden Beeby (1849–1910), who is remembered especially for his later work in the Shetland Isles, was in Scilly and collected *Rumex rupestris* and other specimens now in the herbarium of the South London Botanical Institute.

---

*Crombie named *Ramalina curnowii* for William Curnow. My wife has collected this lichen on Tresco, Nor-Nour, and St. Agnes.

From now on the visits of botanists became more frequent. Ralfs, Tellam, Curnow and James Cunnack (1831–86), all Cornish botanists, made several excursions. In June 1877, the first three were joined by Frederick Janson Hanbury (1851–1938) and Thomas Richard Archer Briggs (1836–91). This company of experienced botanists achieved a number of first records which were mainly published in Ralfs' name. It was at this time that he was compiling the manuscripts which are now in the Morrab Gardens Library, Penzance. These include a small book entitled *Materials towards a Flora of the Scilly Islands* dated 1876, which seems to be a rough note book he carried round on his visits to the islands. From this the records appear to have been copied into Volume III of his manuscript *Flora of West Penrith* dated 1879. I have quoted from this rather than from the earlier note book.

Ralfs was a most able botanist, world famous for his work on diatoms, and in the first rank of local naturalists, but his work on Scilly is disappointing. A sudden failure of eyesight in 1861 had forced him to give up algae in favour of flowering plants and by the time he wrote up his *Flora of West Penrith* he was over seventy. He evidently intended to mark his own records with a '!', but this is so often omitted that it is impossible to distinguish them from unacknowledged quotations from Townsend and others. There are also surprising errors of identification, but in spite of these shortcomings I have been able to credit Ralfs with thirty-four new records from this manuscript.

The Rev Henry Boyden joined the excursion to Scilly arranged by the Penzance Natural History Society on 9 August 1889, and was so impressed that he stayed behind for a few days and repeated the visit. The lists he published contain a number of errors, but he is credited with eight additions to the flora. In July 1890 the islands were visited by Alexander Somerville (1842–1907), a Glasgow botanist, who published a paper in the *Journal of Botany* for April 1893 which claims forty-four new records. Of these, seventeen had been reported previously, thirteen species and five varieties were new, and the rest probably errors. The determinations were made by Arthur Bennett, a leading amateur botanist of the time, and it is surprising to find such unlikely plants as *Vulpia membranacea* in the list. Somerville mentioned the handicap of not having seen the manuscript flora prepared by Ralfs, and this led to the publication by E. D. Marquand a few months later of a paper giving about sixty records from

Scilly (Marquand, 1893). Nearly all of these were taken from Ralfs' *Flora of West Penrith* but there are errors in the transcription and I have preferred to quote direct from Ralfs' manuscript.

In 1906 the islands were visited by William Wright Smith, later Professor Sir William Wright Smith, FRS, (1875–1956) who became the leading Scottish botanist of his generation and concurrently held the three highest appointments at Edinburgh. His visit resulted in two excellent papers. In the first, published in 1907 entitled 'Note on a Peculiar Tussock Formation', he described the extraordinary epiphytic flora on tussocks of a sedge, *Carex paniculata*, in Higher Moors. In the second, published in 1909, he added eight new species to the islands with other useful notes and records.

The publication in 1909 of the *Flora of Cornwall* by Frederick Hamilton Davey (1868–1915) was a most important event in the history of Cornish botany, but unfortunately the treatment of Scilly falls far below the general standard. It seems that Davey never visited the islands himself and he made little attempt to edit the records he collected from various sources, some of which are misquoted. He claimed that the note of exclamation (!) placed after many of the older records indicates that 'either the author or one of his co-workers has seen the plant in the locality mentioned' during the previous six years. In fact these marks are applied fairly consistently to old and otherwise unconfirmed, and sometimes unlikely, records. It seems that Davey used the mark for records lifted from Ralfs' *Flora of West Penrith*, 1879, although these were at least thirty years old. From Davey's *Flora* I have been able to accept seventeen species as recorded from Scilly for the first time on the authority of Tellam, Curnow, C. E. Salmon, F. T. Richards, J. Cunnack and Frank Hill Perrycoste and most of these records are supported by specimens or other evidence.

Table 12   DATES OF KNOWN VISITS UP TO 1940

The following list sets out the visits of botanists and other recorders to Scilly up to 1940. It has been compiled from printed sources, herbarium labels and correspondence but makes no claim to be complete. After 1940 there was a short gap due to the war, and then visitors became frequent with the improvement in travel facilities.

c 1750	Robert Heath—temporarily resident
c 1756	W. Borlase
1813	W. J. Hooker (26 April—5 May)
c 1821	Rev W. T. Bree

Table 12—*continued*

1826	F. K. Eagle
1838	Miss Matilda White (May)
1839	Miss Matilda White
1845	A. J. Hambrough
1849	E. W. Cooke
1850	J. P. Mayne
1852	Misses L. & M. Millett (June, July); J. Ralfs, Joseph Woods (August)
c 1857	J. D. Hooker
1862	F. Townsend (21 June—1 July)
1869	M. A. Lawson
1873	J. Ralfs, R. V. Tellam, W. Curnow (in company, July); W. H. Beeby (July); J. H. Rossall
1876	J. Ralfs & W. Curnow (together in May, Curnow alone in July)
1877	J. Ralfs, T. R. Archer Briggs, F. J. Hanbury, R. V. Tellam, and W. Curnow (all in company in June)
1878	James Cunnack (July)
1883	Rev Augustin Ley (July)
1889	Rev H. Boyden (August)
1890	C. J. Plumtre (March—April); A. Somerville (July); E. S. & C. E. Salmon (end July)
1898	A. G. Gregor (June); F. H. Perrycoste
1902	W. Barratt
1903	Rev H. Boyden (July)
1904	Clement Reid; M. A. Lawson
1905	C. E. Salmon
1906	W. W. Smith (August)
1913	J. W. White; Mrs C. I. Sandwith, N. Y. Sandwith (September)
1921	H. Downes (June)
1922	H. Downes
1925	H. Savage (August)
1932	Miss Margaret Knox (August—September)
1933	Miss E. S. Todd
1934	Miss E. S. Todd (June)
1936	J. E. Lousley (September)
1938	J. E. Lousley (May—June); J. Dallas (June); Miss Margaret Knox (June)
1939	J. D. Grose, J. E. Lousley (July, in company)
1940	J. E. Lousley (May—June)

James Walter White (1846–1932), the Bristol botanist, made two important discoveries in September 1913. A grass, *Calamagrostis epigejos*, was found on the Eastern Isles, and a rush, *Juncus maritimus* var. *atlanticus*, described as new to science from Lower Moors (White, 1914). In 1916 A. J. Hosking discovered Sea Cottonweed,

Otanthus maritimus, of which there was actually an earlier unpublished record. The islands were visited by Harold Downes (1867–1937), a Somersetshire medical man, in 1921 and 1922 who paid special attention to the sedges and grasses.

My own introduction to the Isles of Scilly was in September 1936, when I added *Fumaria occidentalis* to the recorded flora, and was so impressed with the scope for useful work, that it led to a long series of visits covering every month from March until September. By 1940 I had added seventy-seven species to the known flora (Lousley, 1939, 1940, and manuscript). The fortunate chance that the specimens collected by Hambrough, Woods, Beeby and Townsend, and the correspondence and manuscripts of Townsend, were easily accessible to me at the South London Botanical Institute (of which I have been Honorary Curator) suggested the compilation of the present work, the first manuscript of which was completed in 1941.

In recent years many leading botanists have visited the islands and most of them have made some contribution. By far the most numerous have come from J. Donald Grose, author of the *Flora of Wiltshire*, who visited Scilly in 1939 and 1952 and who listed many hundreds of localised records which have been of the greatest help in filling gaps in my own notes of island by island distribution. A valuable detailed study of one island was made by B. W. Ribbons and P. J. Wanstall who stayed on St Agnes in 1948. In 1950 a visit very early in the year by J. E. Raven added *Poa infirma* and *Ophioglossum lusitanicum* to the flora and provided useful notes on other early species. In 1953 Oleg Polunin made the first attempt at classifying the more common plant communities in the islands. Local help has come from Major A. A. Dorrien-Smith, Mrs O. Moyse and Mr and Mrs P. Z. MacKenzie. Their names and those of many visitors are recorded in the following pages.

THE FLORA

This *Flora* is an attempt to collect together all the available information about the flowering plants and ferns found in a wild state in the Isles of Scilly.

Species included in the *List of British Vascular Plants* (Ed J. E. Dandy, 1958) are regarded as accepted members of the British flora. Also included are aliens which have been introduced accidentally, have spread by natural means or, as in the case of windbreaks, are conspicuous, important and long-term features. In Scilly there are many cultivated species deliberately planted out on walls, dunes, etc where they compete for a time with native plants. In general these have no place in a local *Flora*, but it is impossible to be completely logical and a few have been given the benefit of the doubt.

PLAN
SEQUENCE AND NOMENCLATURE
These are based on *List of British Vascular Plants* (Ed J. E. Dandy, 1958) with a few changes to conform with current practice. Entries for species previously reported but not accepted are enclosed in square brackets.

ENGLISH NAMES
Only a few species have Scillonian local names. The English names given are those regarded as in general use.

STATUS
The status refers only to Scilly and is a judgment based on the evidence, often differing from the status in other parts of the British Isles. For example, *Viola odorata* is only a garden escape in Scilly whereas in some parts of England it is regarded as native.

HABITAT
The habitats described are those in which the author or his correspondents have found the species in Scilly and frequently differ from those in which the plants are found on the mainland.

DISTRIBUTION
The units of distribution are islands or groups of islands. These are expressed in the formula:

M.A.T.B.MN. : S.AT.H.TN.E.

in which each letter indicates that the species has been recorded
for the appropriate group, and the absence of a record is indicated
by a dash. The units are as follows:

Inhabited Islands
  M   St Mary's (with Taylor's, Toll's and Newford Islands)
  A   St Agnes (with Gugh)
  T   Tresco
  B   Bryher (with Gweal Island)
  MN   St Martin's (with White, Pernagie and Plumb Islands)
Uninhabited Islands
  S   Samson (with Green, Puffin and White Islands)
  AT   Annet (with Western and Norrad Rocks)
  H   St Helen's (with Round Island, Northwethel and Men-
        a-vaur)
  TN   Tean (with Old Man and Pednbrose)
  E   Eastern Isles

The boundaries are shown on the map on pages 4–5.
This formula shows the distribution of each species at a glance.
It is designed to show especially which plants are restricted to the
inhabited islands and are therefore suspected of being dependent
on human activities or the presence of special habitats, and to show
the extent to which introduced aliens have crossed the sea barriers
from island to island.

FREQUENCY
This is based on the author's field records. It should be noted that
plants may be widely distributed and yet rare, or on the other hand
they may be restricted to one or two islands and yet locally abundant.

FIRST RECORD
This is usually the first *published* record or from a manuscript in a
library (such as Ralfs' *Flora*) but sometimes dated herbarium specimens are the first evidence. These dates are of special importance
in the case of alien plants which have spread since their introduction.

RECORDS
These are listed in the same sequence of islands as the distribution
formula, and are separated by semicolons. Where no recorder's

name is given the author is responsible. Where the record is followed
by a number in brackets, this refers to specimens collected by the
author to support the record (see below).

References to sources in the bibliography are given in the form
'Cooke, 1850' except 'Townsend, 1864' which is quoted so frequently
that it is abbreviated 'Towns.' unless the date is important in the
context.

'!' after a place name indicates that the plant has been seen by the
author in the locality cited. The spellings used by early recorders
are retained.

EXSICCATAE

Flora of the Isles of Scilly. This series of herbarium specimens was
started by the author in 1938 (*J. Bot.* **77,** 196–7, 1939) with printed
labels bearing reference numbers as a basis for this *Flora*. There are
930 collections, and most are represented only in the author's
herbarium; others are also in the collections at Kew and the Natural
History Museum. Only the minimum material necessary for the
purpose was collected and is cited as numbers in brackets following
the records.

Rumices Britannicae Exsiccatae. Numbered specimens collected by
the author to support his work on Docks and Sorrels. These are
cited under *Rumex* in the form '(R.B.E. 270)'.

PERIODICALS CITED

Rep. Bot. Loc. Record Club: *Report of the Botanical Locality Record
    Club* (continued as Botanical Record Club) 1873–86

Rep. B.E.C.: *Report of the Botanical Exchange Club of the British
    Isles* (later Botanical Society and Exchange Club of the British
    Isles) 1879–1948

Watsonia: *Watsonia*, the Journal of the Botanical Society of the
    British Isles, 1949→

Proc. B.S.B.I.: *Proceedings of the Botanical Society of the British
    Isles*, 1954–69

Rep. Watson B.E.C.: *Report of the Watson Botanical Exchange
    Club*, 1884–1934

Scillonian: *The Scillonian*, 1925→. The local magazine published
    at St Mary's. Quoted from the run in the British Museum
    Library.

HERBARIA CITED

Herb. Beeby: Collection of William Hadden Beeby (1849–1910) in Herb. S.L.B.I. (SLBI)

Herb. Curnow: Collection of William Curnow (1809?–87) transferred from Imperial College of Technology to British Museum (Nat. Hist.). (BM).

Herb. Druce: Collection of George Claridge Druce (1850–1932) at Department of Botany, University of Oxford (OXF).

Herb. D-S: Collection of wildflowers from Scilly made by Miss Gwen Dorrien-Smith and Miss Ann Dorrien-Smith (now Mrs Phillimore) between 1922 and 1940. An unreliable list appears in Vyvyan, 1953, 154–160. Cited from the specimens at Tresco Abbey.

Herb. Downes: Part of the collection of Harold Downes (1867–1937) now at Yeovil Museum (YEO). Efforts to trace the remainder have been unsuccessful.

Herb. Univ. Glasgow: Herbarium of the Department of Botany, The University of Glasgow (GL). Includes plants collected by A. Somerville (1842–1907).

Herb. Edinburgh: Herbarium of the Royal Botanic Garden, Edinburgh (E).

Herb. Kew: Herbarium of the Royal Botanic Gardens, Kew (K).

Herb. S.L.B.I.: Herbarium of the South London Botanical Institute, 323 Norwood Road, SE24, London (SLBI).

Herb. Tellam: Part of the collection of Richard Vercoe Tellam (1826–1908). The specimens seen were brought up from the Museum boiler room and the remainder had probably been destroyed. County Museum and Art Gallery, Truro (TRU).

Herb. Thurston: Collection of Edgar Thurston (1855–1935) in Herb. Kew (K).

Herb. Townsend: Collection of Frederick Townsend (1822–1905) in Herb. S.L.B.I. (SLBI).

Herb. Woods: Collection of Joseph Woods (1776–1864) in Herb. S.L.B.I. (SLBI).

PTERIDOPHYTA

LYCOPSIDA
SELAGINELLACEAE
Selaginella Beauv.
S. kraussiana (Kunze) A. Braun.   Mossy Clubmoss.
This creeping plant, so common in greenhouses on the mainland, is thoroughly established in the lower part of Tresco Abbey Gardens (723, det Alston), where I first noted it in 1954. A native of tropical and southern Africa and the Azores, it is established in Cornwall and the milder parts of Ireland.

SPHENOPSIDA
EQUISETACEAE
Equisetum L.
E. arvense L.   Field Horse-tail.   Native.
Very rare   M.-.-.-.-. : -.-.-.-.-.   First record: this *Flora*.
St Mary's: In quantity in potatoes in the fifth field up from Old Town Church, 1959. Miss B. M. C. Morgan.

PTEROPSIDA
OSMUNDACEAE
Osmunda L.
O. regalis L.   Royal Fern.   Native.
Peaty places in the marshes, and wet places below sea cliffs. Local
M.-.T.-.-. : -.-.-.-.-.   First record: Cooke, 1850.
St Mary's: Marsh near Old Town, Cooke; abundant in Heugh Town and Old Town Marshes, Towns.—it is still plentiful in many places in Higher and Lower Moors; shore east of Porth Minick; shore at Tregear's Porth. Tresco: West of Tresco Pool, Towns.— still plentiful in swamp at west end of Great Pool.

DENNSTAEDTIACEAE
Pteridium Scop.
P. aquilinum (L.) Kuhn (*Pteris aquilina* L.). Bracken. Native.
Abundant on heaths, cliff slopes, sandy waste, etc. wherever the soil is sufficiently deep for the rhizomes to ramify and exposure to wind and salt spray not excessive.
M.A.T.B.MN. : S.AT.H.TN.E.   First record: Cooke, 1850.

91

ADIANTACEAE

Adiantum L.

A. capillus-veneris L.   Maidenhair Fern.   Native.

Damp hollow in 'Head' on sea-cliff.   Very rare.

M.-.-.-.-. : -.-.-.-.-.   First record: Lousley, 1967.

St Mary's: Tregear's Porth (849), 1959. This persisted for several years but in 1967 a fall of cliff destroyed the habitat.

BLECHNACEAE

Blechnum L.

B. spicant (L.) Roth   Hard Fern.   Native.

In a pit on a heath.   Extinct.   M.-.-.-.-. : -.-.-.-.-.

First record: Millett, 1852—as *Blechnum boreale*.

St Mary's: in a pit on Salakee Down, near the Giant's Castle, Millett, 1852. Davey (1909) marks Millett's record as recently confirmed, but Ralfs failed to find it in 1876, and many other botanists have searched without success.

B. chilense (Kaulf.) Mett.   Established alien.

In a malodorous ditch under elms.   Very rare.

M.-.-.-.-. : -.-.-.-.-.   First record: Lousley, 1967.

St Mary's: Ditch by the lane leading from Salakee to Porthellick Pool, 1953, O. Polunin (Polunin 21, det. Alston); 1959 (839); 1963 (841); still increasing, 1967. This fern is a native of Chile and belongs to a group of very closely allied species from New Zealand, South Africa and South America which are grown in greenhouses.

ASPLENIACEAE

Phyllitis Hill

P. scolopendrium (L.) Newm.   (*Scolopendrium vulgare* Sm.)

Hartstongue.   Native.

Damp banks, walls and wells.   Rather rare.

M.A.T.-.MN. : -.-.-.-.-.   First record: Cooke, 1850.

Scattered records from all the inhabited islands except Bryher, but usually in small quantity.

Asplenium L.

A. adiantum-nigrum L. ssp. *adiantum-nigrum*.   Black Spleenwort.

Native.

Hedgebanks and walls.   Uncommon.

M.A.T.B.MN. : S.-.-.-.-. First record: Cooke, 1850.
Distributed about the six largest islands and especially on St Mary's,
but less frequent than *A. billotii* with which it is often confused.

A. billotii F. W. Schultz (*A. obovatum* auct., *A. lanceolatum* Huds.).
Lanceolate Spleenwort.   Native.
Walls of fields and buildings.
M.A.-.B.MN. : -.-.-.-.-. Common.   First record: Bree, 1831.
This fern is common on St Mary's and St Agnes, local on Bryher
and St Martin's and strangely unrecorded for Tresco. The fronds
remain green through the winter and may be found fresh and
bearing spores in spring. They shrivel in dry weather in early
summer and are often so dried up by July and August that they
may be difficult to find.

A. marinum L.   Sea Spleenwort.   Native.
Rocks, walls and cliffs, especially by the sea, but also on buildings
and high rocks a little inland, and up to 100 ft above sea level.
Abundant.   (Plate p 193).
M.A.T.B.MN. : S.AT.H.TN.E.   First record: Bree, 1831.
Cooke (1850) describes this species as 'the grand botanical feature
of Scilly', and it certainly occurs in surprising size and abundance.
He found fronds nearly 33 in long, and I have found them 30 in
in length, though commonly they are about half this. Some of the
largest specimens have long parallel-sided pinnules ('var'. *parallelum*
Moore).

A. trichomanes ssp. *quadrivalens* D. E. Meyer.   Maidenhair
Spleenwort.   Native.
Old walls.   Rare.   M.-.T.-.MN. : -.-.-.-.-.   First record:
Tonkin, 1893.
St Mary's: Walls of old mill on Garrison, Tonkin, 1893; garden wall,
Hugh Town, 1953, Polunin. Tresco: plentiful on walls of green-
houses outside Gardens, 1959, B. M. C. Morgan. St Martin's: wall,
Upper Town (875 det. A. C. Jermy).
A. ruta-muraria L. Wall-rue
Cooke in 1850 wrote 'I have also an impression that *A. ruta-muraria*
was growing on a stone wall in St Mary's between the town and Holy
Vale'. This has not been confirmed.

Ceterach DC.
C. officinarum DC.   Rustyback.   Native.

Old wall.   Very rare.   -.-.-.-.MN. : -.-.-.-.-.
First record: Lousley, 1967.   St Martin's: Higher Town, 1956.
Although so abundant in many parts of south-west England, this
fern is very rare in the West Penrith area of Cornwall.

ATHYRIACEAE

Athyrium Roth
A. filix-femina (L.) Roth.   Lady Fern.   Native.
Marshes, ditches and damp hollows.   Local.
M.-.T.-.-. : -.-.-.-.-.   First record: Cooke, 1850.
St Mary's: Old Town Marshes, Cooke, 1850; marsh between
Hugh Town and Old Town, and Higher Moors near Porthellick,
Tonkin, 1893; Old Town Marsh and Green, Grose; I have seen it
in all these localities and also at Salakee and in plenty in Watermills
Lane. Tresco: Townsend, 1864; in hollows on Castle Down.

ASPIDIACEAE

Dryopteris Adans.
D. filix-mas (L.) Schott.   Male Fern.   Native.
Hedgebanks, ditches and under rocks.   Frequent.
M.A.B.T.MN. : S.-.-.-.-.   First record: Cooke, 1850.
Widely distributed in St Mary's but more local on the other large
islands.

D. borreri Newm.   Borrer's Male Fern.   Native.
Laneside banks.   Rare.   M.-.-.-.-. : -.-.-.-.-.
First record: Lousley, 1967.
St Mary's: Laneside near Telegraph (748, teste Alston); Porthloo
Lane (749, teste Alston). The fronds of this species tend to be
tougher than those of *D. filix-mas* and often overwinter.

D. carthusiana (Vill.) H. P. Fuchs (*Lastrea spinulosa* C. Presl).
Narrow Buckler Fern.   Native.
Rare.   M.-.T.B.-. : -.-.-.-.-.   First record: Cooke, 1850.
St Mary's: Old Town Marsh, Cooke. Tresco: withy bed east of
Tresco Pool, Townsend, 1864. Bryher: under big rock beyond
Charlies' en route to pools, 1937, Herb. D.-S.

Fig 6 1 *Lavatera cretica*. 2 *Muehlenbeckia complexa* (leaf x2). 3 *Allium babingtonii*. 4 *Cyrtomium falcatum* (indusium x5)

1

2

3

4

D. dilatata (Hoffm.) A. Gray.   Common Buckler Fern.   Native.
Damp and shady places.   Frequent.
M.-.T.B.-. : S.-.-.-.-.   First record: Cooke, 1850.
The apparent absence from St Agnes and St Martin's is explained
by the scarcity of damp and shady habitats on these islands.

D. aemula (Ait.) Kuntze   Hay-scented Buckler Fern.   Native.
'Marsh'.   Very rare.   M.-.-.-.-. : -.-.-.-.-.
First record: Cooke, 1850 (as *Lastrea recurva*).
St Mary's: 'One locality', Cooke, 1850; Marsh between Hugh
Town and Old Town, Tonkin, 1893. Also given for St Mary's by
Ralfs, 1879, and for Scilly by Millett, 1852.
Although efforts to refind this attractive fern have been unsuccess-
ful, there is no reason to doubt these records. It is plentiful near
Penzance so Ralfs and the Misses Millett would know it, and
Cooke was a most accurate recorder. It is not a marsh species but
may yet be refound on a damp sheltered hedgebank in Old Town
Marsh.

Polystichum Roth
P. setiferum (Forsk.) Woynar   Soft Shield Fern.   Native.
Probably a shady hedgebank.   Very rare.   M.-.-.-.-. : -.-.-.-.-.
First record: Cooke, 1850.   St Mary's: Ralfs, 1879.

P. aculeatum (L.) Roth   Prickly Shield Fern.   Native.
Probably a shady hedgebank.   Very rare.
Recorded by Cooke, 1850, without locality. The records for both
Shield Ferns have not been confirmed in recent years but there is
no reason to doubt their accuracy.

Cyrtomium C. Presl
C. falcatum (L.f.) C. Presl   Established alien.
Very rare.   Under boulders on shore.   (Fig. 6, p 95)
M.-.-.-.-. : -.-.-.-.-.   First record: Lousley, 1967.
St Mary's: under boulders on shore below cliff east of Porth
Minick, June 1956, Mr and Mrs R. C. L. Howitt. Fronds collected
by me two months later (753) were named by A. H. G. Alston
There were two clumps of this fern growing with *Agrostis stoloni-
fera*, *Beta maritima*, *Crithmum maritimum*, *Hypochoeris radicata*,
Osmunda regalis and *Rumex crispus*. The habitat is a very natural

ANT COMMUNITIES ON SAND: (*top*) unconsolidated dune with Marram Grass, *Ammophila naria*, dominant at Bar Point, St Mary's; (*bottom*) consolidated dune joining granite hillocks at Appletree Banks, Tresco

one but household rubbish has been tipped into the sea on the opposite side of the small bay, and it is likely that roots were washed up from this source. Alternatively spores may have blown from plants in cottage gardens at Old Town, where it is cultivated. It has persisted, and in 1967 I found two clumps ten yards apart which I think were not the original ones.

C. *falcatum* is a native of southern Japan, Korea, Formosa and eastern China and commonly cultivated. It has been recorded as an escape in Alabama, Florida and Hawaii, and is also found in the Azores. It has been recently recorded from two places in Holland (van Ooststroom in *Ned. Kruidk. Archf.*, **57**, 217–8, 1950).

THELYPTERIDACEAE

Thelypteris Schmidel

T. palustris Schott This was entered, as *Lastrea thelypteris*, by C. J. Plumtre in his note book as found in St Mary's in April, 1890. There were then a few wet peaty habitats where the plant might have grown but the record is best treated as an error.

POLYPODIACEAE

Polypodium L.

P. vulgare L. *sensu lato.* Common Polypody. Native.
Walls, rocks and hedgebanks. Common.
M.A.T.B.MN. : S.-.-.TN.E. First record: Cooke, 1850.
This is now divided into cytotypes which have been given sub-species status. Two of these occur in Scilly:
ssp. *vulgare* (*P. vulgare* L. *sensu stricto*). Tresco: Wall in Abbey Gardens (846 det A. C. Jermy); Abbey Gardens, 1962, J. F. M. Cannon, Herb. Mus. Brit.
ssp. *prionodes* Rothm. (*P. interjectum* Shivas). St Mary's: old wall, Tremelethen (661, det. Jermy). Tresco: Monument Hill (885, det. J. A. Crabbe). Bryher: Near Look-out Hill (888, det. Crabbe). St Martin's: Upper Town (877, det. Crabbe). This is the common plant.

OPHIOGLOSSACEAE

Botrychium Sw.

B. lunaria (L.) Sw. Moonwort. Native.
Sandy ground, growing under bracken. Very rare.
M.-.-.-.MN. : -.-.-.-.-.-. First record: Millett, 1852—without locality.

St Mary's: Bar Point in the neighbourhood of the telegraph, Millett in Marquand, 1893. Still in the same place, 1940. St Martin's: Tonkin, 1893. In the south of England the appearance of Moonwort is irregular and, in spite of much recent disturbance in the vicinity of the cable, it is quite likely to reappear.

Ophioglossum L.
O. vulgatum L.   Adder's Tongue.   Native.
This, the common Adder's Tongue of the mainland ($2n = 480$) appears to be very rare in Scilly. The only material I have seen was from St Agnes, collected by J. E. Dallas in a damp area at the north end of Wingletang Bay, 13 June 1938. The fronds were very much larger than those of the next species.

O. azoricum C. Presl (*O. vulgatum* ssp. *polyphyllum* (A.Br.) E. F. Warburg; var. *polyphyllum* A.Br.; var. *ambiguum* Coss and Germ.). Native.
Peaty sand, often under bracken and with bluebells, and in heathy places.   Locally abundant.   M.A.T.-.MN. : -.-.-.-.-.
First record: Millett, 1852 (from Bar Point, see Marquand, 1893). St Mary's: Bar Point, Millett; Bar Point (605); rough fields near Watermills, April 1928, Herb. D.-S. St Agnes: On heights to the north-east, Townsend, 1864, *B.E.C. Rep.* for **1870**, 18–19, 1871, and Herb. Townsend. Still there and on dunes near Priglis Bay, 1967, Miss H. M. Quick (884). Tresco: exposed hillside, R. W. J. Smart, Herb. Kew; in several places about Appletree Banks (617) St Martiñ's: sandy soil north and west of Middle Town, Townsend, 1864; common in many places on the western half of this island and repeatedly collected from here by Curnow, Cunnack, Tellam, Ralfs and others; Porth Seal (432).
It has recently been suggested that *A. azoricum* with $2n=720$ is a late Tertiary alloploid derivative of the 32-ploid *O. vulgatum* and the 16-ploid *O. lusitanicum*. It occurs from the Azores up the western coast of Europe to the Orkneys (Löve, A. and Kapoor, B. M., 1967, *Svensk bot. Tidskr.*, **67**, 29–32 and *Nucleus*, **9**, 132–138).

O. lusitanicum L.   Native.
Shallow peaty soil over granite   Very rare. (Fig 7, p 125).
-.A.-.-.-. : -.-.-.-.-.   First record: Raven, 1950.
St Agnes: rough heath in the southern half of the island, Raven, 1950; Lousley, 1953 (612), D. McClintock and B. T. Ward.

O. lusitanicum has been known in Guernsey since 1854 and several attempts have been made to give this name to the Scillonian *O. azoricum*, which also occurs in the Channel Isles. The relationship was discussed by W. Roberts in 1884 (*Science Gossip*, **20**, 148–9), and Ralfs used the name in 1877 (*in. litt.* to Townsend), and it was still being used by Tonkin in 1893 (p 113). It was not until March 1950 that the true plant was found by Mr John Raven. The reason it was overlooked for so long is that the fronds are active during the winter and turn yellow and disintegrate in early April before most botanists arrive.

The main colony of about a hundred fronds covers just over a square metre at the foot of a large boulder. The soil is peaty and shallow and rather bare of vegetation. I noted the following associated species in April 1953: *Sedum anglicum*—ab, *Plantago coronopus*—ab, *Calluna vulgaris*, very stunted—f, *Anagallis arvensis*—r, *Ulex* sp.—two seedlings.

O. lusitanicum ($2n=240$) occurs in the Channel Isles and in Brittany south to Spain and the Mediterranean.

SPERMATOPHYTA

GYMNOSPERMAE

PINACEAE

Pinus L.

P. radiata D. Don (*P. insignis* Douglas).   Monterey Pine.
Planted alien.
Shelter belts.   Common.   M.A.T.-.MN. : -.-.-.-.
This native of California is one of the few Pines with the needles
arranged in three's. It is commonly planted as a windbreak on
account of its rapid growth and wind-resisting qualities.

P. contorta Douglas.   Lodge Pole Pine.
Since 1964 this has been planted instead of *P. radiata* and to
replace some of the old windbreaks. It is readily distinguished by
the arrangement of the needles in pairs

CUPRESSACEAE

Cupressus L.

C. macrocarpa Hartweg.   The Monterey Cypress.   Planted
alien.
This tree proved especially valuable in providing early shelter for
Tresco Gardens—'The sea-living Cypress is the conifer that
succeeds best on the islands' (Carmichael, 1884). It is still used
occasionally and good examples are to be seen in the plantation
at Gimble Porth, Tresco (520).
Various other conifers are also planted on Tresco.

ANGIOSPERMAE

DICOTYLEDONES
RANUNCULACEAE

Caltha L.

C. palustris L.   Marsh Marigold.   Introduced.
This grows in two places near Tresco Abbey. Here it is planted as an ornamental plant. As a native it seems to be absent from Scilly as it is also from the Channel Isles.

Clematis L.

C. vitalba L.   Travellers' Joy.   Introduced.
Recorded by Mr and Mrs R. C. L. Howitt from by Holy Vale Guest House, St Mary's in 1960, this must be regarded as a garden escape, as it is in the Channel Isles.

Ranunculus L.

R. acris L.   Meadow Buttercup.   Native.
Damp meadow and wet trackside.   Very rare.
M.A.T.-.-. : -.-.-.-.-.-.   First record: Ralfs, 1879.
St Mary's: Ralfs' *Flora;* in a marshy meadow near Rocky Hill. Tresco: by Pool Road. St Agnes: Miss H. M. Quick. In both stations the plant grows in much wetter ground than is usual in most parts of England but in Cornwall marshy meadows are the general habitat.

R. repens L.   Creeping Buttercup.   Native.
Roadsides, field borders, marshes and gardens.   Common.
M.A.T.B.MN. : -.-.H.TN.-.   First record: Townsend, 1864.

R. bulbosus L.   Bulbous Buttercup.   Native.
Sandy ground.   Common.
M.A.T.B.MN. : S.-.-.TN.-.   First record: Townsend, 1864.

R. arvensis L.   Corn Buttercup.   Colonist.
Cultivated ground.   Very rare.   First record: Lousley, 1967.
St Mary's: the garden of the Star Castle Hotel, 1940.

R. muricatus L. Prickly-fruited Buttercup.   Colonist.
Weed in bulbfields and gardens, waste places.   Abundant.
(Fig 8, p 155).
M.A.T.B.-. : -.-.-.-.-.   First record: Downes, 1924.
St Mary's: 'in quantity and established', 1923, Downes in *Rep.
B.E.C.*, **7**, 165 and 372, 1924; 'a curse in St Mary's', A. A. Dorrien-
Smith *in litt*, 1938; by 1939 it was abundant in bulbfields in the
southern half of the island, including those on the Garrison,
towards Peninnis, and about Old Town; it has now spread to
Pelistry Bay and the north of the island. St Agnes: Middle Town,
above Cove Vean, etc, 1953. Tresco: bulbfields, New Grimsby,
1953; Northward (770), 1957; etc. Bryher: Pool (557) and Southward, 1939. I have failed to find it on St Martin's.
As a native this species is widespread in the countries round the
Mediterranean, and it is now naturalised on the east and west
coasts of North America and in Australia and New Zealand.
In Britain it occurs as a casual introduced with grain, and sometimes wool, and it was established near Prestwick, Lancashire,
from 1875 to 1893 (*Rep. B.E.C.*, **1**, 398, 1894). It had almost
certainly been in Scilly for some years before Downes noticed it in
quantity in 1923, and the climate and system of bulb cultivation are
so well suited to its life history that it is likely to become a permanent feature of the flora. *R. muricatus* commences to flower as
early as March, and fruits and dies off in May and June. The fruits
have therefore dropped by the time the bulbfields can be cleaned
or cleared, and its growing season coincides with that of the crop.
In 1951 it was found in fields at Poltesco on the Lizard, where
it may also find conditions suitable.

R. sardous Crantz   Hairy Buttercup.   Native.
Sandy flats and weed in bulbfields.   Uncommon.
M.A.-.B.-. : -.-.-.TN.-.   First record: Townsend, 1864.
St Mary's: 'near the freshwater pools', Townsend; near Holy Vale,
Grose; bulbfield near Trenoweth; abundant and 2 ft tall in bulbfield
near the Telegraph, 1963. St Agnes: near freshwater pools, Towns-
end; Middle Town, Grose; by Priglis Pool (687); Love Lane, 1967.
Bryher: Herb.D.-S. Tean: a few plants, Dallas.
Somerville (1893) recorded var. *parvulus* (L.) Rouy and Fouc.,
without locality, but this was probably only a small starved
form.

R. marginatus D'Urv. var. *trachycarpus* Fisch. and Mey.   Colonist.
Abundant weed in arable fields.   Very local. (Fig 8, p 155).
First record: Lousley, 1955 (see below).
St Martin's: in April 1950 Dr R. C. L. Burges collected material
in flower from near Higher Town which it was not possible to name
at the time. In 1952, J. D. Grose found a single plant on the roadside
a little beyond the Post Office, which had immature fruits and in
April 1953 I took roots from arable fields by Pounds Lane which
I grew on in my garden (573, 635) to obtain fruits for identification
(*Proc. B.S.B.I.*, **1,** 463, 1955).
The St Martin's Buttercup is restricted to a few fields at Upper
Town but in these it is in such abundance that when in flower in
May the fields are coloured yellow and are conspicuous from a
distance. It is a native of the Near and Middle East, and is found
in Britain occasionally as a grain alien. It resembles *R. sardous* but
the carpels have tubercles over the whole surface of the disk, and
are terminated at the apex by a conspicuous triangular or lanceolate
beak at least 1 mm long, and sometimes slightly curved.

R. parviflorus L.   Small-flowered Buttercup.   Native.
Bulbfields, gardens, roadsides and sandy banks.   Common.
M.A.T.B.MN. : -.-.-.-.-.   First record: Townsend, 1864.
This species is now common and even locally abundant in all the
inhabited islands with the possible exception of St Agnes, where I
have noticed it only at Middle Town. Townsend gave it as rare
and recorded it only from 'near Star Castle, St Mary's and St Agnes',
while Ralfs (1879) gave it as 'very scarce' from St Mary's only.
W. H. Beeby collected it in 1872 from 'turfy places, The Hugh'
(Herb. S.L.B.I.) and this was probably a native station and the one
known to Townsend and Ralfs. It still grows under similar conditions at, for example, Southward Bay, Bryher, but the usual
habitat now is bulbfields. It seems that this is a native species with a
life-cycle well suited to the system of cultivation of the bulbfields,
and it has spread from natural habits and increased to its present
abundance since the introduction of the flower industry.

R. flammula L. subsp. *flammula*.   Lesser Spearwort.   Native.
Marshes and margins of pools.   Local.   (Map 1, p 105).
M.-.T.B.-. : -.-.-.-.-.   First record: Townsend, 1864.
St Mary's: Boggy field near Hugh Town, Anne Dorrien-Smith in

Ranunculus flammula, Lesser Spearwort

Thurston, 1929; Lower Moors, Grose; common in Higher and Lower Moors. Tresco: common about the pools; margin of Abbey Pool (8; 515; 725). Bryher: Townsend, 1862 in Herb. Townsend; by the large pool; in a splash below Timmy's Hill.

This species exhibits a very wide range of variation in Scilly. At one extreme is the creeping form with narrow-lanceolate leaves, usually less than 2 mm wide, and small flowers seldom as much as 1 cm in diameter, which is plentiful on the wet sandy margin of Abbey Pool. Townsend (1864) recorded *R. reptans* L. from the freshwater pools on St Mary's and Tresco, and no doubt had this form in mind. It is represented in his herbarium by specimens labelled as collected from Bryher, and these are similar to plants found in dune-slacks all round the coast of the mainland. These are a distinct looking phenotype which Padmore has called forma *tenuifolius* (Wallr.) Padmore (Padmore in *Watsonia*, **4,** 19–27, 1957) At the other extreme in Scilly are relatively luxuriant plants with broad lower leaves which grow in wet fields in St Mary's and Tresco. The records of Miss Anne Dorrien-Smith and J. D. Grose were given as 'var. *serratus* DC.'. Such forms are common in Higher and Lower Moors but do not attain the size of robust plants sometimes found on the mainland.

R. hederaceus L.   Ivy-leaved Crowfoot.   Native.
Inland pools, splashes in the marshes, ditches and roadside gutters.
M.-.-.-.-. : -.-.-.-.-.   Rare, local and decreasing.
First record: Millett, 1852.
St Mary's: Old Town, 1862, Towns. in Herb. Towns.; common in the marshes, Towns., 1864; Holy Vale, Grose; Watermills Pond; Higher Moors (412); roadside gutter, Watermills (7); roadside, Tremelethen—the last two, a small form; Old Town Marsh; Porthellick, 1955, Howitt.
R. omiophyllus Ten. (*R. lenormandi* F. W. Schultz). Error. This species was recorded by Somerville (1893) without locality, and by the writer (1939). My own plants proved to be a minute form of *R. hederaceus* with leaves seldom as much as 5 mm across at their greatest width. Such plants are easy to misidentify and further evidence is required before the species can be accepted.

R tripartitus DC. (*R. lutarius* (Revel) Bouvet).   Native.
Extinct.   In the acid part of Higher Moors.
M.-.-.-.-. : -.-.-.-.-.   First record: Ralfs, 1879.
In a letter to Townsend in 1876 Ralfs reported finding *R. intermedius* in 'High Marsh [sic] St Mary's', and the record is repeated in his manuscript *Flora*. It is supported by a specimen labelled 'Higher Bog, St Mary's, small flowered and a good fruiter, 26 May, 1876' collected by John Ralfs and now in the herbarium of Imperial College, London. It was collected again by F. J. Hanbury on 30 June, 1887 (Herb. Mus. Brit.).

R. trichophyllus Chaix   Fennel-leaved Water Crowfoot.
Native.
In a pond.   -.-.T.-.-. : -.-.-.-.-.   Very rare.
First record: Davey, 1909 (but see below).

This species is included on the evidence of a specimen from Tresco labelled 'Near bottom duckery gate' (Herb. D.-S.). I have found only *R. baudotii* in this locality.

R. trichophyllus subsp. *drouetii* (Godr.) Clapham is given for Scilly on the authority of Tellam in Davey, 1909. Confirmation is needed.

R. aquatilis L. has been recorded by Ralfs and others but all the specimens so labelled which I have seen are *R. baudotii*, which still occurs in forms with large floating leaves in all the localities cited.

R. baudotii Godr.   Native.
Brackish pools.   Common.
M.A.T.N.MN. : -.-.-.-.-.   First record: Townsend, 1864.

R. baudotii is common and probably occurs in every pool or pond throughout the islands. So much salt spray is driven over the land in gales that even the larger 'freshwater pools' contain appreciable amounts of salt. Although the most characteristic habitat is heavily brackish pools on the coast, such as those on Bryher and St Agnes, the species is just as much at home in Scilly in roadside inland pools such as the one at Newford, St Mary's.

The species is exceedingly variable, ranging from a drawn up state for which Hiern proposed the name of *R. thalassois*, and others with submerged leaves only, to robust specimens with large floating leaves including some referrred to *R. confusus* Godr. The fine series of specimens collected in Scilly by Ralfs and Curnow and cultivated material grown from them, now in the herbarium of Imperial College, is convincing evidence of the plasticity of *R. baudotii*.

R. ficaria L.   Lesser Celandine.   Native.
Under bracken on the cliff slopes and in hedges.   Abundant.
M.A.T.B.MN. : S.AT.H.TN.E.   First record: Ralfs in Marquand, 1893.

This species dies down early in Scilly and hence was overlooked by the early observers. The common plant is the diploid subsp. *ficaria* which has no axillary bulbils. In addition, subsp. *ficariiformis* Rouy and Fouc. occurs in St Mary's: between Hugh Town and Old Town (585) and near Higher Moors. This is much larger, about 25 cm tall, with leaves up to 5.5 x 5.5 cm with rather large pointed bulbils in their axils. In sprawling habit and general appearance it recalls the luxuriant Jersey plant which Moss called forma *luxurians*.

Aquilegia L.

A. vulgaris L.   Columbine.   Garden escape.

This was found on the edge of the *Osmunda* swamp, Porthellick, St Mary's by Miss B. M. C. Morgan in 1959 where I suspect that it has spread from the garden at Salakee. On Tresco, Miss M. B. Gerrans found it in 1963 'amongst rocks on sand-dunes near the Abbey'. She regarded it as 'not native' and her specimen is of a garden columbine (Herb. Mus. Brit.). The species is native and common in Cornwall.

Thalictrum L.

T. minus L. subsp. *minus*.   Lesser Meadow-rue.   Doubtful native.
On sandy cliff.   M.-.-.-.-. : -.-.-.-.-.   Very rare.
First record: Lousley, 1967.

St Mary's: a single patch on a sandy cliff below the school, Carn Thomas, August 1956. I first treated this as a garden escape although the habitat closely resembles the places where the species grows on the Lizard. However, the leaves lack the glands of the garden plant and the recent detection of *Salvia horminoides*, another Lizard plant, a few yards away, suggest that it is more likely to be an overlooked native.

PAPAVERACEAE

Papaver L.

P. rhoeas L.   Common Poppy.   Colonist.
Weed of cultivation.   Rare.
M.A.T.-.MN. : -.-.-.-.-.   First record: Millett, 1852.

St Mary's: Cornfields, very scarce, Ralfs; bulbfields, Halangy; abundant in one field near Old Town; bulbfield near Porth Minick. St Agnes: Miss H. M. Quick. Tresco: a single plant in Abbey Gardens. St Martin's: High Town, Grose.

Ralfs knew Scilly before bulbfields replaced cornfields and it seems that on these light and often sandy soils *P. rhoeas* has always been scarce.

P. dubium L.   Long-headed Poppy.   Colonist.
Weed of cultivation.   Common.
M.A.T.B.MN. : -.-.-.-.-.   First record: Townsend, 1864.

This is the common poppy of bulbfields, gardens and other cultivated ground.

P. lecoqii Lamotte This is recorded in Ralfs' *Flora* from St Agnes on the authority of Tellam. Downes wrote 'It seems to be the prevailing poppy in the Scilly Isles, 1921' (in Thurston and Vigurs, 1922a, p 6), and he noted in his copy of the *Flora of Cornwall* that he found it on St Mary's and St Martin's. I have seen no plants from Scilly which I would refer to *P. lecoqii* as found in the Eastern Counties, and regard the records as errors for *P. dubium*.

P. somniferum L.   Opium Poppy.   Established alien.
Weed in cultivated fields and on a rubbish tip.   Very rare.
M.-.T.-.MN. : -.-.-.-.-.   First record: Smith, 1909.
St Mary's: bulbfield by Old Town Lane; Pelistry Bay, on a rubbish tip. Tresco, occasional, Smith. St Martin's: a few plants in a bulbfield near Upper Town, Grose.

P. orientale. L.   Oriental Poppy.
This garden outcast was reported by Grose in 1952 from Old Town Bay, St Mary's.

Glaucium Mill.
    G. flavum Crantz   Yellow Horned Poppy.   Native.
Sand and shingle shores.   Very common.
M.A.T.B.MN. : S.-.-.TN.E.   First record: Borlase, 1756.

FUMARIACEAE

Fumaria L.
All my exsiccatae of this genus were determined by the late H. W. Pugsley.

F. occidentalis Pugsl.   Western Fumitory.   Native.
Weed in bulbfields and on walls.   Locally abundant. (Fig 8, p 155; Map 2, p 110).
M.-.-.-.-. : -.-.-.-.-.   First record: Lousley in *Rep. B.E.C.*, **11**, 238, 1937
St Mary's: When first found this handsome fumitory was abundant in bulbfields and on walls within the area bounded by Old Town Bay (22; 803), Parting Carn, and Salakee (684)—see *J. Bot.*, **77**, 197, 1939. It has extended its ground very slightly and was observed in a garden near Hugh Town Church in 1952, and in a field a little beyond Parting Carn in 1953. This very slow rate of spread during the thirty years I have known it would seem to indicate that it had

Fumaria occidentalis, Western Fumitory

already been in the Old Town district for a long time when first discovered. On the other hand it seems incredible that such a conspicuous plant could be overlooked by earlier visitors and particularly so as it is to be found in flower in one field or another throughout the summer. It is an endemic species known elsewhere only in Cornwall.

F. capreolata L.   Ramping Fumitory.   Native.
Walls and rocks.   Rather rare.
M.A.T.-.MN. : -.-.-.-.E.   First record: Millett, 1852.
St Mary's: Ralfs, 1877 in Herb. Curnow; between Hugh Town and Old Town; High Cross Lane; hedge by Garrison path; wall by Peninnis Road. St Agnes: below Parsonage; Old Lane. Tresco:

New Grimsby, Grose. St Martin's: near St Martin's Bay, Grose.
The following specimens have been identified by Pugsley as var.
babingtonii Pugsl.: St Mary's: 1877 and 1878, W. Curnow in Herb.
Mus. Brit.; roadside wall near Old Town (21). Great Ganinick:
South End, Dallas (18)—see *J. Bot.*, **77**, 197, 1939.

F. bastardii Bor.   Colonist.
Weed in bulbfields.   Frequent.
M.A.T.B.MN. : -.-.-.-.-.   First record: Townsend, 1864
(as *F. confusa*).
Frequent throughout the larger islands. The following have been
identified by Pugsley:
var. *bastardii:* St Agnes: bulbfield near the Church (26); Tresco:
Old Grimsby (14).
var. *hibernica* Pugsl.: Scilly, Townsend (see Davey, *J. Roy. Inst.
Cornw.*, **52**, 7, 1905). St Agnes: near the Church (26a, 388). St
Martin's: Middle Town (27, 27a).
var. *gussonei* Pugsl. St Mary's: London (25a). Bryher: under
Timmy's Hill (16, 17).

F. muralis Sond. ex Koch subsp. *boraei* (Jord.) Pugsl.   Colonist.
Weed in bulbfields and on walls.   Common
M.A.T.B.MN. : -.-.-.-.-.   First record: Somerville, 1893.
First evidence: St Mary's 1860–61, Herb. Mus. Brit.
St Mary's: Old Town ('probably var. *britannica*'); bulbfield near
Old Town (24); roadside wall near Old Town (20); near Holy Vale
(19, 'the typical species'). St Agnes: frequent. Tresco; bulbfield,
Old Grimsby (15, 'the typical species'); Dolphinstown (789).
Bryher: Southward, Grose. St Martin's: bulbfield, Middle Town (28).
var. *britannica* Pugsl. St Mary's: near Porth Cressa Bay, Salmon,
1905 det. Pugsley (Herb. Mus. Brit.)—recorded as var. *serotina*
Clavaud in Davey, 1909.

F. officinalis L.   Common Fumitory.   Colonist.
Weed in cultivated ground.   Frequent.
M.A.T.N.MN. : -.-.-.-.-.   First record: Millett, 1852.

CRUCIFERAE
Brassica L.
B. rapa L. (*B. campestris* L.)   Turnip.   Escape from cultivation.

Reported from St Agnes: bulbfield, 1948, Ribbons and Wanstall, and St Mary's: near Holy Vale, as an annual in bulbfield.

B. nigra (L.) Koch   Black Mustard.   Probably introduced. Waste places near the shore.   Very local.
M.-.-.-.-. : -.-.-.--.   First record: Townsend, 1864.
Known only from the neighbourhood of Hugh Town, St Mary's, where it was locally plentiful but is now decreasing. This species is accepted as native in Cornwall but its restriction in Scilly to the neighbourhood of the only town suggests an old introduction.

Sinapis L.
S. arvensis L. (*Brassica arvensis* (L.) Rabenh.).   Charlock. Colonist.
Weed in cultivated ground.   Frequent.
M.A.T.B.MN. : -.-.-.-.-.   First record: Millett, 1852.
Occurs with glabrous and also with hairy fruits.

S. alba L. (*Brassica alba* (L.) Rabenh.)   White Mustard.   Relic of cultivation.
Cultivated ground.   Rare.
-.A.T.-.-. : -.-.-.-.-.   First record: Lawson, 1870.
St Agnes: Lawson, 1870; as a crop in two fields, 1953. Tresco: garden of New Inn and nearby fields, 1952; Back Lane, 1967.

Diplotaxis DC.
D. muralis (L.) DC.   Sand Rocket.   Established alien.
Sandy waste ground and walls.   Rare.   M.-.T.-.-. : -.-.-.-.-.
First record: Downes in Thurston and Vigurs, 1922a, 15.
St Mary's: Hugh Town, 1952, Grose and Lousley; Porthcressa (589—f. *caulescens* Kit.). Tresco: Downes, as above; on a wall, Herb.D.-S.; New Grimsby (372).
In St Mary's this has spread since it was first found in 1952 in The Park, Hugh Town to walls in the vicinity. In Tresco I have found it only at New Grimsby and if this is the place where Downes collected it in 1921, the species shows no evidence of rapid spread in Scilly.

D. tenuifolia (L.) DC.   Wall Rocket.   Established alien.
Walls and trackside.   Very rare.   -.-.T.-.-. : -.-.-.-.-.

First record: Tresco, rare. Smith, 1909.

Tresco: as above, near Abbey, 1938.

In contrast to the rapid spread in south-east England, Wall Rocket in Scilly has not extended its range in sixty years.

Raphanus L.

R. raphanistrum L.   Wild Radish.   Colonist.

Cultivated fields.   Rather rare.   M.A.T.-.-. : -.-.-.-.-.

First record: Somerville, 1893 (without locality).

St Mary's: 1938, M. Knox; Star Castle Hill, Grose; near Tremelethen. St Agnes: Middle Town, Grose; The Gugh (petals white), Grose; Lower Town. Tresco: Old Grimsby.

var. *aureus* Wilmott (flowers yellow with concolorous veins) is becoming increasingly frequent on St Mary's: Holy Vale (39), Lousley 1939, 198; Salakee, Polunin; Parting Carn.

R. maritimus Sm.   Sea Radish.   Native.

Sandy shores and a wall-top.   Local. (Map 3, p 114).

M.A.-.-.-. : -.-.-.-.-.   First record: Townsend, 1864.

St Mary's: 1878, W. Curnow, Herb. Essex; Porth Cressa; Pelistry Bay; The Garrison (38); Porth Minick; Porth Mellin. St Agnes: Priglis Bay.

Plants with white flowers (var. *albus* Druce) occur at Porth Cressa in addition to the usual yellow-flowered form.

R. sativus L.   Garden Radish.   Casual on sandy waste ground.

Very rare.   First record: Vyvyan, 1953.

Tresco: Near Abbey Farm, 1928. Herb. D.-S.

Crambe L.

C. maritima L.   Sea Kale.   Native.

On the drift-line on shingly beaches, and 'in a field'.   Rather rare.

(Map 4, p 115).   M.A.T.B.MN. : -.-.-.-.E.

First record: Found in 1898, Perrycoste in Davey, 1909.

St Mary's: 'One plant in a field at St Mary's, 1898, probably an escape', Perrycoste, l.c.; St Mary's, Salmon, 1905, in Davey, 1909; Old Town shore, 1906, Smith 1909; 5 plants at Porth Minick. 1938; one plant at Porthellick, 1938. St Agnes: Smith, 1909; in fair quantity at Priglis Bay, 1938; Porth Coose, 1952, Grose. Tresco: 1906, Smith 1909; in great plenty, 1920, A. J. Hosking in

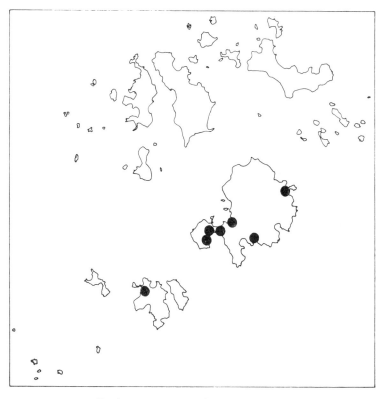

Raphanus maritimus, Sea Radish

Downes' copy of Davey; Appletree Banks, Herb. D.-S.; New Grimsby, 1939, Grose. Bryher: one plant, West Bay, 1952, A. Conolly; two plants, 1967; many 1970. St Martin's: 1 plant at Porth Seal, 1940. Eastern Isles: Great Ganilly, 1939.

The recorded history of this species in Scilly goes back only to 1898 when it was found by Perrycoste in a habitat where it was unlikely to have been native. If the plant was present earlier it seems remarkable that it escaped the keen eyes of the Misses Millett, Townsend, Beeby, Ralfs and others. In the nineteenth century Sea Kale was grown extensively as a market crop in Scilly (Brewer, 1890) and this no doubt explains Mrs Perrycoste's record. Salmon in 1905, and Smith in 1906, expressed no doubts about status and although

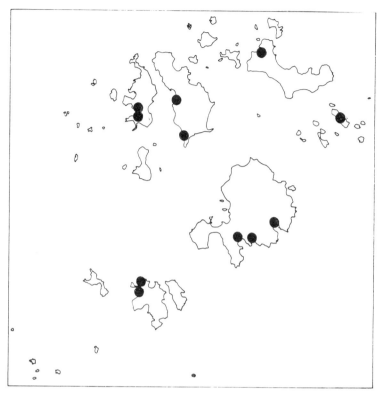

Crambe maritima, Sea Kale

at Old Town Bay it could have originated from a crop, the species is more likely to have arrived as a seed carried by ocean currents. It is widespread on the drift-line of the coast of western and northern Europe and subject to sudden local population increases. In Scilly there was a period of spread between 1905 and 1940.

Cakile Mill.
C. maritima Scop.　　Sea Rocket.　　Native.
Sandy shores.　　Rather rare.
M.A.T.B.MN. : -.-.-.-.E.　　First record: Townsend, 1864.
St Mary's: Porth Cressa, Grose; Old Town Beach; Bar Point; inland, near Buzza Hill on a heap of seaweed for use as manure; Porthloo, Halangy Point, Dallas. St Agnes: near Gugh Farm;

Porth Coose, Grose. Tresco: Appletree Banks, Grose; Gimble Porth; Pentle Bay. Bryher: beach near Samson Hill; Great Porth. St Martin's: New Quay, Grose; Great Bay; near Old Quay. Great Ganilly, MacKenzie.

Lepidium L.
L. campestre (L.) R.Br.   Reported by 'B.P.' in *Science Gossip*, **20**, 99, 1884, but probably an error for the next species.
L. heterophyllum Benth. (*L. smithii* Hook.).   Smith's Cress. Native.
Waste ground and roadsides.   Rare.   M.-.T.-.-. : -.-.-.-.-.
First record: Lawson, 1870 (without locality).
St Mary's: London, Grose; near Coastguard Station, Grose (3951). Tresco: Ralfs, 1879.

Coronopus Zinn
C. squamatus (Forsk.) Aschers. (*C. procumbens* Gilib.)   Wart Cress.   Native.
Waste places, gateways to fields and trampled roadsides.   Rather rare.   M.A.T.B.MN. : -.-.-.-.-.   First record: Millett, 1852.
St Mary's: frequent, Grose; between Hugh Town and Old Town. St Agnes: near Lighthouse. Tresco: Abbey Grounds, Grose; New Grimsby. Bryher: near Pool, Grose; west side, near cottages. St Martin's: one place on south-west coast, Grose; Pool Green. Although so widely distributed, this species is far less frequent than the next.

C. didymus (L.) Sm.   Lesser Swine Cress.   Colonist.
Waste and cultivated ground, roadsides.   Common.
M.A.T.N.MN. : -.-.-.-.-.   First record: Townsend, 1864.
A native of South America, first recorded in Britain from Cornwall in 1778, this species is now common in Scilly as in many other parts of southern England.

Thlaspi L.
T. arvense L.   Penny Cress.   Colonist.
Waste and cultivated ground, roadsides.   Common.
M.A.T.B.MN. : -.-.-.-.-.   First record: Townsend, 1864.
Apparently less common in St Martin's than in the other four inhabited islands.

C ipsella Medic.

C. bursa-pastoris (L.) Medic.   Shepherd's Purse.   Native.
Weed in cultivated land, and waste places.   Very common.
M.A.T.B.MN. : -.-.-.-.-.   First record: Millett, 1852.

Cochlearia L.

C. officinalis L.   Scurvy-grass.   Native.
Maritime rocks and shingle.   Common.
M.A.T.B.MN. : S.AT.H.TN.E.   First record: Townsend, 1864.
This fleshy species thrives on rocks heavily exposed to salt spray.
While common in such places all round the coast of the larger
islands, it is at its best on smaller ones exposed to the Atlantic
where it is one of the few species able to survive. On sea-swept rocks
of the Western Isles Scurvy-grass is exceptionally large and fleshy—
from Melledgen and Rosevear I have seen leaves attaining a
width of 6.5 cm.

C. danica L.   Danish Scurvy-grass.   Native.
Rocks, walls and shingle and not confined to the coast.   Very
common.
M.A.T.B.MN. : S.AT.H.TN.E.   First record: Millett, 1852.
In no other part of the British Isles have I seen this species as
abundant and ubiquitous as it is in Scilly. A minute state in which
the whole fruiting plant measured only about 1 x 1 cm occurred in
the crevices of the masonry of Cromwell's Castle, but has now
been lost through repointing. In April 1957 Miss M. Jaques
collected a plant near Turfy Hill, St Martin's which has characters
of both *C. danica* and *C. officinalis*, and may be a hybrid.

Ionopsidium (DC.) Reichb.

I. acaule (Desf.) Reichb.   My attention was drawn to this by the late
A. A. Dorrien-Smith as established as a weed on paths in Abbey Gardens,
Tresco (360, 362). A native of Portugal, with an odour of honey, it
superficially resembles *Cochlearia danica*.

Aurinia

A. saxatile (L.) Desv. (*Alyssum saxatile* L.)   'Tresco, established,
May 1929, R. Meinertzhagen' (Herb. Mus. Brit.). This specimen looks
like a luxuriant garden-grown plant.

Lobularia Desv.

L. maritima (L.) Desv.   Sweet Alyssum.   Established alien.

Old walls and waste ground.   Locally common.

M.-.T.-.MN. : -.-.-.-.-.   First record: Townsend, 1864.

St Mary's: one specimen on shore near Heugh Town, Townsend, 1864; now common on walls and waste ground in and around Hugh Town. Tresco: dunes, Old Grimsby, 1957. St Martin's: south coast, Grose, 1952; below school, 1963.

This garden escape has a continuous history in Scilly going back to Townsend's single plant a century ago. By 1875 Curnow found it on walls and it probably reached its maximum abundance about 50 years ago, since when building has reduced the sandy ground available round Hugh Town. The species is native on shores of the Mediterranean and is well established in the Channel Isles and on the English coast.

Erophila DC.

E. verna (L.) Chevall.   Whitlow-grass.

This was recorded from St Mary's by Lawson, 1870 but this early flowering species would have been long over at the time of his visit. Davey, 1909 cites 'Scilly Isles, Townsend' but this was an error of transcription. The species cannot be accepted for Scilly without confirmation.

Armoracia Gilib.

A. rusticana Gaertn., Mey., and Scherb. (*Cochlearia armoracia* L.). Horse Radish.   Naturalised alien.

Waste places.   Very rare.

-.-.T.-.-. : -.-.-.-.-.   First record: Smith, 1909.

Tresco, without locality, Smith; near School, Herb. D.-S.; still there, 1957.

Cardamine L.

C. pratensis L.   Cuckoo Flower, Lady's Smock.   Native. Marshes and ditchsides.   Local.

M.-.T.-.-. : -.-.-.-.-.   First record: Millett, 1852.

St Mary's: Higher Moors! and Upper Watermills Lane!, Ralfs' *Flora;* Lower Moors; Old Town Marsh, Howitt; below Salakee towards Porthellick (632), Mrs O. R. Moyse. Tresco: near the Withy Bed, Tresco Abbey.

Mr D. E. Allen has examined the specimen from Porthellick (632) and writes '*C. fragilis* (Lloyd) Bor. Resembles the usual English form and not that of Cornwall and north-west France.'

C. flexuosa With.   Wavy Bitter-cress.   Native.
Damp shaded ground.   Rare.   M.-.T.-.-. : -.-.-.-.
First record: Ralfs, 1879—as *C. sylvatica* Link.
St Mary's: Ralfs in litt. to Townsend, 1876. Tresco: Abbey Grounds
(33).

C. hirsuta L.   Hairy Bittercress.   Native.
Damp places and field borders.   Uncommon.
M.A.T.-.MN. : -.-.-.-. First record: Townsend, 1864.
St Mary's: Porthellick Pool; Near Rocky Hill; Watermill Cove, etc.
St Agnes: Near Cove Vean. Tresco: Ralfs *Flora*. St Martin's:
sandy track below Lower Town (566); bulbfields, Higher Town
Bay.

Barbarea R.Br.
B. vulgaris R.Br.   Common Winter-cress.   Established alien.
Disturbed ground and on a wall.   Very rare.
-.-.T.-.-. : -.-.-.-. First record: Boyden, 1889.
Tresco: on wall north of Tresco Church, and Duckery, Herb. D.-S.;
kitchen garden of the Abbey. Also recorded from Scilly by Tellam
in Davey, 1909, but this record, like Boyden's, is unlocalised.
One of the specimens in the Dorrien-Smith herbarium was named
at Edinburgh as *B. arcuata* (J. and C. Presl) Reichb.

B. verna (Mill.) Aschers.   Land Cress.   American Land-cress.
Colonist.
Roadsides.   Rare.   M.-.T.-.MN. : -.-.-.-.
First record: Ralfs, 1877 (in litt. to Townsend).
St Mary's: near Gateway to Garrison, Miss B. M. C. Morgan.
Tresco: near Old Grimsby, Ralfs, *Flora;* still there—on School
Green.—see also Lousley, 1940. St Martin's: towards Lower
Town, Herb. Dallas; Lower Town (841), plentiful, 1959.
Land Cress is grown as a salad and was formerly more popular
than it is today. It is believed to be native in the western Mediter-
ranean region, Canaries, etc., but is widely naturalised. It has
persisted at Old Grimsby for 90 years and at Lower Town, St
Martin's for at least 25.
B. intermedia Bor. is reported by R. P. Bowman as found in 1957 on
St Mary's 'in waste places along the road after leaving Hugh Town for
Rocky Hill' but no specimen is available.

Nasturtium R.Br.

N. officinale R.Br.   Watercress.   Native.

Ditches, streamside, roadside gutters, and ponds.   Local.

M.-.T.-.-. : -.-.-.-.-.   First record: Millett, 1852.

St Mary's: Higher and Lower Moors; Old Town Marsh (651); Salakee, Grose; Watermill Lane (491); ponds at Newford (762). Tresco: Ralfs, *Flora;* Well Cover, by Abbey Gardens, R. Lancaster. I have not seen any specimens which could be referred to *N. microphyllum* (Boenn.) Reichb.

Rorippa amphibia (L.) Bess. This was listed by Grose, 1939, from Old Town Marsh and near Porthloo, St Mary's but has not been confirmed and appears to be an error.

Matthiola R.Br.

M. incana (L.) R.Br.   Garden Stock.   Garden escape established.   Walls and sandy shores.   Rare.

M.A.T.-.-. : -.-.-.-.-.   First record: Lousley, 1967.

St Mary's: walls, etc. 1875, W. Curnow, Herb. Curnow; Porth Cressa, 1940; shore, Old Town, 1953. St Agnes: Higher Town, well established, 1970, R. M. Burton. Tresco: established on shore, Old Grimsby, 1953 onwards. Stocks are commonly grown in gardens in Scilly.

M. sinuata (L.) R.Br. This rare native species was recorded without locality by Millett, 1852 and Montgomery, 1854. Ralfs made the following entry in his *Flora:*—'Searched all the islands unsuccessfully in 1872. J.R. It is probably become extinct.' From its distribution in Britain there is no reason why the species should not have occurred, but it is much more likely that *M. incana* was the plant found.

Sisymbrium L.

S. officinale (L.) Scop.   Hedge Mustard.   Native.

Roadsides, waste places and edges of bulbfields.   Abundant.

M.A.T.B.MN. : -.-.-.-.-.   First record: Millett, 1852.

S. orientale L.   Eastern Rocket.   Naturalised alien.

Sandy waste ground.   Rare.   M.A.T.-.-. : -.-.-.-.-.

First record: Downes in Thurston and Vigurs, 1922a.

St Mary's: Star Castle Hill, 1939, Grose; Porth Cressa, 1940 (420). St Agnes: By Gugh Farm, 1939 (32). Tresco: Downes, as above; New Grimsby, 1938 (30).

This alien, a native of countries round the Mediterranean, has shown little tendency to spread in Scilly since 1940.

Arabidopsis (DC.) Heynh.
A. thaliana (L.) Heynh.   Thale Cress.   Native?
Weed in cultivated ground, and sandy waste.   Very rare.
M.-.T.-.MN. : -.-.-.-.-.   First record: Lousley, 1967.
St Mary's: bulbfield, Halangy, 1940 (460). Tresco: weed in Abbey Gardens, 1953. St Martin's: East end of Higher Town Bay on sandy waste ground, 1953; by Perpitch, 1967.
Although this species is accepted as native almost throughout the British Isles, its recent discovery and rarity in Scilly suggests that here it is a recent arrival brought in perhaps with bulbs or seed for cultivation.

Lunaria L.
L. annua L.   Honesty.   Established garden escape.
M.-.T.-.MN. : -.-.-.-.-.   First record: Lousley, 1967.
St Mary's: naturalised, 1967, P. Z. MacKenzie. St Agnes: on slope below track from jetty, Porth Conger, 1959; naturalised there, 1963, W. H. Hardaker; in plenty in bracken and brambles on slope not far from cottage, 1967. St Martin's: Higher Town Bay, 1971.

RESEDACEAE

Reseda L.
R. luteola L.   Dyer's Rocket.   Native—probably extinct.
M.-.-.-.-. : -.-.-.-.-.   First record: Millett, 1852.
St Mary's: given for this island by Lawson, 1870 and Ralfs, 1879.

R. lutea L.   Wild Mignonette.   Native?
Sandy waste ground, and in a bulbfield.   Very rare.
M.A.-.-.-. : -.-.-.-.-.   First record: Townsend, 1864.
St Mary's: rare, Townsend, 1864. St Agnes: waste ground and in a bulbfield near Gugh Farm. This has not been refound on St Mary's, and the locality on St Agnes is one where several aliens have appeared.

R. alba L.   White Mignonette.   Established alien.
Sandy waste ground.   Now very rare and decreasing.
M.-.-.-.-.-. : -.-.-.-.-.   First record: Millett, 1852.

First evidence: F. K. Eagle, 1826 in Herb. Mus. Brit.
St Mary's: Common on waste ground around Hugh Town, Townsend, 1864—as var. *suffruticulosa* L.; sandy ground near the sea,
1872, Beeby in Herb. S.L.B.I.; Hugh Town, plentiful, 1873, J. H.
Rossall in Herb. Mus. Brit.; sandbanks and waysides, very common,
1878, W. Curnow in Herb. Mus. Brit.; Porthmellin, 1923, Downes
in Herb. S.L.B.I., Herb Kew, and *Rep. B.E.C.*, **7,** 375, 1924; Hugh
Town, 1938 onwards; near Harry's Walls, 1959.
There are many other records which add nothing to those given
above except to confirm that White Mignonette has a continuous
history about Hugh Town extending for over 140 years. It may
have originated as a garden escape, as Townsend assumed, or have
been brought in by shipping from the Mediterranean, but it was
abundant on the sandy isthmus on which Hugh Town has been
built. Much of the ground is now built on, and the last large
colony known to me was destroyed by the dumping of road metal
in the spring of 1939. Since 1952 it has become exceedingly scarce.

VIOLACEAE

Viola L.

V. odorata L.   Sweet Violet.   Garden escape.
M.-.T.-.-. : -.-.-.-.-.   First record: Lousley, 1967.
St Mary's: a large patch by Garrison Walk near Morning Point
Tresco: Back Lane (white-flowered), 1970, R. M. Burton.

V. riviniana Reichb.   Dark Wood Violet.   Native.
Hedgebanks, downs and dunes under bracken.   Abundant.
M.A.T.B.MN. : S.AT.H.TN.E.
First record: Townsend, 1864 (as '*V. sylvatica* Fries—Common': a
dwarfed form frequent on the downs and in exposed situations.
This is the common violet of Scilly. The plants are often condensed
and very floriferous and then fall under subsp. *minor* (Gregory
Valentine. Flowering is by no means confined to the spring and
seldom visit Scilly without finding the species in bloom—in August
1956 flowers were abundant.

V. reichenbachiana Jord. ex Bor.   Pale Wood Violet.   Native
Hedgebanks and under bracken.   Well distributed but rare.
M.A.T.-.MN. : S.AT.H.-.E.   First record: Lousley, 1967.
St Mary's: Tremelethen; Innisidgen; Pelistry. St Agnes: Gugh

Tresco: Appletree Banks; Merchant's Point. St Martin's: White
Island. Samson. Annet. St Helen's. Great Ganinick, Dallas,
Herb. Dallas. Middle Arthur.

These dark-spurred violets are not identical with the species as I
know it in woods and hedgebanks in less exposed places in south-
east England. They appear to be restricted in Scilly to places where
comminuted shells of mollusca in the sand provide calcium car-
bonate. In Devon *V. reichenbachiana* is generally distributed
but it is rare in Cornwall where acid soils prevail.

V. canina L. subsp. *canina*   Dog Violet.   Native.
Heathy cliffs and sand dunes.   M.A.T.-.-. : -.-.-.-.-.
First record: Millett, 1852, but this was in the aggregate sense and
the first evidence of the species as now understood is a specimen
dated 1930 in Herb. Dorrien-Smith.
St Mary's: Old Town Churchyard; Golf Course; Bar Point (472
det. P. M. Hall as var. *pusilla* Bab.; 594); frequent in the north of
the island; Bar Point, Grose (3951, det. P. M. Hall as var. *ericetorum*
Reichb.). St Agnes: Gugh; Love Lane. Tresco: Appletree Banks to
Monument Hill; Appletree Banks (616); 'edge of pool, May 10,
1930', Herb. D.-S.; near Cromwell's Castle (455, det. P. M. Hall as
var. *pusilla* Bab. 'a narrow-leaved form similar to plants received
from Spurn Head incorrectly named *V. lactea*.').

x *riviniana*—Persistent at Bar Point, St Mary', where I have
collected it at intervals (458 and 471 teste P. M. Hall, 593). A large
fertile plant from St Agnes, under gorse at St Warna's Bay (434)
was thought by P. M. Hall to be possibly this hybrid if not a shade
form of *v. canina*.

V. lactea Sm. This was recorded for Scilly by J. Jacob in his *West Devon
and Cornwall Flora*, 1835–37, and the record has been repeated in later
works. Ralfs writing to Townsend in 1876 claimed to have found it and
gave it for St Mary's in his *Flora*. There are suitable habitats, and it is
locally common in Cornwall, but it has been searched for repeatedly
and it is likely that the old records were based on forms of *V. canina* with
which it is frequently confused.

V. tricolor L. subsp. *tricolor*   Heartsease.   Native.
Cultivated ground.   Rare.   M.-.-.-.MN. : -.-.-.-.-.
First record: Millett, 1852 (in the aggregate sense): Smith, 1909
(who also gives 'var. *arvensis*'.).
St Mary's: Smith, 1909; bulbfield by shore, Pelistry Bay (600).

St Martin's: bulbfields, Higher Town Bay (571); undated, R. V. Tellam in Herb. Tellam as *V. curtisii* Forst. var *symei;* some of the other old records of *V. curtisii* from the fields below the School-house on St Martin's also belong here (see below).

V. arvensis Murr.   Field Pansy.   Native.
Cultivated ground.   Locally common.
M.-.T.-.MN. : -.-.-.-.-.   First record: Smith, 1909.
This was formerly divided into a number of segregates now regarded as of no taxonomic value. All seven of my gatherings were determined by the late P. M. Hall as *V. segetalis* forma *obtusifolius* (Jord.) Drabble, while other segregates which have been recorded include *V. ruralis* Jord., *V. agrestis* Jord., and *V. deseglisei* Jord.

V. kitaibeliana Schult. (*V. nana* (DC.) Godr.)   Dwarf Pansy.
Native.
Sand dunes and sandy cultivated land.   Rare. (Fig 7, p 125).
M.-.T.B.MN. : -.-.-.TN.-.   First record: Curnow in *Hardwicke's Science Gossip*, **1876,** 162 as *V. curtisii.*
The species was discovered on Tresco by J. Ralfs in July 1873 and named by him as *V. curtisii* var. *mackaii* in a letter to Townsend in May 1876. It was recognised as *V. nana* Corb. by E. G. Baker in 1901 (*J. Bot.*, **39**, 11), but in 1881 C. C. Babington had placed it in the correct group when he wrote 'Small form from Scilly is very like *V. parvula* Tin.' (*Manual*, ed. 8, p 44).
St Mary's: sandy field near shore, Pelistry Bay (634, 667). It per-sisted here in decreasing numbers from 1953 to 1957 accompanied by puzzling plants with the flowers and small fruits of *kitabeliana* but nearer *arvensis* in habit and stipule form. Tresco: New Grimsby, Ralfs, 1873 (see above, but he wrote 'Old Grimsby' in error), where it was collected at intervals from sandhills until the site was buried under the foundations of the air-station in the 1914–18 war; since 1936 it has been seen frequently by Tutin, Raven and others on and about Appletree Banks (701) in the south of the island. Bryher: south valley, April 16, 1951, Herb. D.-S. Seen over a small

Fig 7 Native Rarities: 1 *Ornithopus pinnatus* (fruit x1½). 2 *Ophioglos-sum lusitanicum* (enlargement x2). 3 *Trifolium occidentale*. 4 *Juncus maritimus* var. *atlanticus* (spikelet x5). 5 *Viola kitaibeliana* (flower x2½). 6 *Poa infirma* (spikelet x3, persistent glumes x3)

2

3

4

5

6

area almost annually since. St Martin's: sandy field, June 1877, Ralfs, Herb. Kew. Seen there, below the school, at intervals by many botanists since. Tean: West Porth, plentiful, 1953 (623).

V. kitaibeliana commences to flower at the end of March and then, and during April, it is a distinct little plant not easily confused with other species. In natural habitats on sand dunes it is usually seen in disturbed places, such as round rabbit holes, but it can also grow in fairly close dune turf. In such places it dries up and disappears early, but in arable ground it can persist as straggling late-flowering plants into June or even early July. The early collectors found mainly these untypical plants but, even so, it is difficult to understand how they came to name them *V. curtisii* or *V. mackaii*, which are large flowered pansies with the petals longer than the sepals.

V. kitaibeliana occurs on the Mediterranean coast, the western coasts of France and the Channel Isles. Until recently the Isles of Scilly were the northernmost locality but it was found in 1956 near Ambleteuse, Pas-de-Calais (Lousley in *Le Monde des Plantes*, **319**, 19, 1956). The plant of Scilly appears to be identical with that of Jersey and material from Tresco and Bryher has been found to have the same chromosome number, $2n=24$ (Pettet in *Watsonia*, **6**, 41, 1964).

POLYGALACEAE

Polygala L.

P. vulgaris L.   Common Milkwort.   Native.

Downs, cliffs, sandy and heathy places.   Common.

M.A.T.B.MN. : -.-.-.-.-.   First record: Millett, 1852.

P. dubia Bellynck (*P. oxyptera* Reichb.) Specimens from Tresco. Tean and Bar Point, St Mary's have been so named, but Glendinning has shown that this species cannot be maintained (*Proc. B.S.B.I.*, **1**, 93, 1954).

P. serpyllifolia Hose   Heath Milkwort.   Native.

Heathy places.   Common.

M.A.T.B.MN. : S.-.-.-.E.   First record: Somerville, 1893.

P. calcarea F. W. Schultz. Recorded by Townsend, 1864 from St Martin's, but in his own copy of the paper he altered the name to *P. depressa* Wend. (i.e. *P. serpyllifolia*) with a reference to a specimen in his herbarium which can no longer be found. Undoubtedly an error—see also Davey, 1909 p 63.

Hypericum pulchrum, Upright St John's Wort

GUTTIFERAE

Hypericum L.

H. inodorum Mill. (*H. elatum* Ait.).   Tall Tutsan.   Established alien.

Hedgebank.   Very rare.

M.-.-.-.-. : -.-.-.-.   First record: Lousley, 1967.

St Mary's: hedgebank 'near Porthellick', 1939, Grose (3833). The locality is at the north end of Old Town Lane, nearly half a mile from Porthellick, and I saw it there in 1940, well established and not near a house: Howitt, 1960.

H. humifusum L.   Trailing St John's Wort.   Native.

Walls, roadside gutters and heathy places.   Frequent.

M.A.T.B.MN. : -.-.-.-.   First record: Townsend, 1864.

var. *ambiguum* Gillot has been claimed for Scilly by Miss E. S. Todd, but I have no further information.

H. pulchrum L.   Upright St John's Wort.   Native.

Rough ground on cliff slopes.   Rare. (Map 5 p 127).

M.-.-.-.-. : -.-.-.TN.E.   First record: Ralfs, 1879.

St Mary's: Ralfs' *Flora;* Star Castle Hill, Grose; above Bar Point, Grose; Innisidgen Point. Tean: Grose. Eastern Isles: North Hill, Ganilly, in brambles near the landing beach, 1935, Herb. D.-S.; Great Arthur, very local (316).

The restricted distribution is difficult to explain but it seems to be limited to places which have never been cultivated.

H. elodes L.   Bog St John's Wort.   Native.

Growing in spongy bog.   Very rare.

M.-.-.-.-. : -.-.-.-.-.

First record: Millett, 1852, but there is a specimen from St Mary's collected by A. Hambrugh in 1845 in Herb. Townsend.

St Mary's: Higher Moors—limited to a small bog near Tremelethen (274), decreasing, and exceedingly scarce by 1967.

PITTOSPORACEAE

Pittosporum Banks ex Soland.

P. crassifolium Soland. ex Putterl.   Karo.   Naturalised alien.

Granite carns, walls and amongst boulders on shores. (Plate p 257).

Common and increasing.   M.A.T.B.MN. : -.-.-.TN.-.

First record: as self-sown, Vyvyan, 1953, but many earlier notices as commonly planted, e.g. Lousley, 1939, 234. (Fig 9, p 185).

Karo is now so common as self-sown plants that there is no object in giving detailed localities. It is most common growing from cracks in the granite of carns round the coast, and amongst boulders along the shores and on cliffs, but it also occurs inland on walls by lanes and round fields. It has even been found growing from a grating in the main street of Hugh Town (Vyvyan, 1953).

The history of *P. crassifolium* in Scilly is of considerable interest. It is a native of the Kermadec Islands and of North Island, New Zealand, where it grows on the coast as a shrub 15–30 ft tall, and was first grown in Scilly in Abbey Gardens. The specimen illustrated in the *Botanical Magazine, t.* 5978, was provided from here in 1872. It proved remarkably wind-hardy, rapid and erect-growing, and stood up well to salt spray and summer droughts. Karo proved an excellent windbreak under the exacting local conditions, and was planted throughout the inhabited islands to provide shelter for the bulbfields.

Until 1952 it was known to me only as a planted shrub, but in that year many young plants were found growing on rocky granite carns, and in the following year I found them on rocky beaches. The plants on the carns were all about 5–6 ft tall, and in positions which the seed is unlikely to have reached unless dropped by a bird (Plate p 257). Since the original sudden outbreak many further plants have been found.

The most probable explanation of the sudden widespread appearance of self-sown plants is that many of the planted windbreaks reached maturity at about the same time. The dark red flowers appear in April, and are succeeded by tomentose subglobose fruits, about an inch in diameter, with three woody valves, which split to reveal numerous black seeds. These it seems are eaten by a bird which voids or otherwise deposits them on carns and beaches. The bird has not yet been identified but some of the habitats, including one on Tean, are up to a kilometre from the nearest planted shrubs.

A fungus, *Rosellinia necatrix* (Hart.) Berlese, which causes bulb rot in Narcissus also causes root rot in *Pittosporum crassifolium*.

P. ralphii T. Kirk. A closely allied shrub from New Zealand, with the leaves abruptly narrowed to the petiole with flat instead of revolute margins, and the capsules smaller (*c* 1.5 instead of 2.5 cm long), is said to be also planted. It, and intermediates with *P. crassifolium* which may be hybrids, are grown at Tresco Abbey.

MYRTACEAE

Leptospermum J. R. and G. Forst.

L. scoparium J. R. and G. Forst.   Manuka.   Established alien.
Tresco: 'in places *Leptospermum scoparium* had spread by seeding to a remarkable extent, and clusters of shrubs of all sizes in full flower strongly resembled a growth of Blackthorn in April', E. Brown, 1935.; thoroughly established with abundant seedlings in Abbey Wood, above the garden wall, 1963. A native of New Zealand.

L. pubescens Lam. This Australian species was seen in Abbey Wood in 1963 growing with the above and equally well established.

PLANT COMMUNITIES OF EXPOSED HABITATS: (*top*) Thrift, *Armeria maritima*, dominant on An
(*bottom*) Ling, *Calluna vulgaris*, and Gorse, *Ulex europaeus*, on thin soil over granite. Win
tang Downs, St Agnes

Leptospermums have been grown in the open without protection at Tresco Abbey for a century (*Gdnrs'. Chron.*, **1872**, 1129.).

Eucalyptus L'Hérit.
E. iinearis Dehnh.   Established alien.
Tresco: Abbey Wood, 1963. Freely established from self sown seed above the Abbey Garden wall. A native of Tasmania.

TAMARICACEAE

Tamarix L.
T. gallica L.   French Tamarisk.   Naturalised alien.
Hedges, about buildings and sandy wastes.   Common.
M.A.T.B.MN. : S.-.-.-.-.   First record: Bree in Forbes, 1821 (and see below).
A native of the western Mediterranean, this shrub was one of the first to be planted as a windbreak in Scilly. It was planted extensively in the early part of the nineteenth century and impressed many early visitors including W. J. Hooker who noted it at Holy Vale (in litt to Dawson Turner, April 27, 1813). It now occurs commonly in hedgerows and in most of the places where it is apparently wild, as for example about the ruins on Samson, it persists from former planting for shelter. In some places such as Lower Town, St Agnes there are plants with very thick trunks which must be old. Occasionally it appears to be self-sown as, for example, at Old Town Bay, St Mary's (497) and near Pool, Bryher (522).

T. africana Poir.   African Tamarisk.   Established alien.
Tresco: Old Grimsby, in hedgerows close to the sea, 1963, M. B. Gerrans, ref. 1148, det. B. Baum (Herb. Mus. Brit.).
The species of this genus are difficult to distinguish and it may be that several occur.

ELATINACEAE

Elatine L.
E. hexandra (Lapierre) DC.   Waterwort.   Native.
On sandy mud on the margin of a slightly brackish pool.   Rare.
-.-.T.-.-. : -.-.-.-.-.   First record: Townsend, 1864.
Tresco: 'margin of Tresco pool near Mr Smith's house', Townsend, 1864. There are specimens in Herb. Dorrien-Smith from 'edge of pool near Abbey Green', and it was seen here by Downes in 1921

and 1922 (Herb. Kew), and I found it in abundance in 1939 (369), 1952, 1956 (724), 1957 and subsequently. The flowering time extends from June to September, and in recent years it has been so plentiful that the margin of Abbey Pool has been coloured a rich red from the leaves and flowers of this small plant. The pool has become less brackish and this, with the close cropping and droppings of the ornamental wildfowl, may account for the increase.

CARYOPHYLLACEAE

Silene L.

S. vulgaris (Moench) Garcke (*S. cucubalus* Wibel). Bladder Campion.
This was recorded by Millett, 1852 (as *S. inflata*) and Ralfs' *Flora* cites T. Millett as having found it on St Agnes. Lawson, 1870 claims to have found it on St Mary's. The Milletts were generally reliable, but they do not give the closely related *S. maritima* which is plentiful on St Agnes. Lawson's 'Additions' includes a number of errors. The species cannot be accepted without confirmation.

S. maritima With.   Sea Campion.   Native.
Rocks, walls and maritime cliffs.   Frequent.
M.A.T.B.MN. : S.-.H.TN.E.   First record: Townsend, 1864.
Sea Campion occurs in scattered colonies on all the inhabited islands, and on Samson and Puffin Island, Tean, St Helens, Middle Arthur, Great Ganinick and Little Ganilly. On Samson there are two adjacent patches; one with the calices uniformly pale yellow (43a) and the other with them flushed with purple which is darkest on the veins and teeth (43). These have remained constant since I first noticed them in June 1938.

S. gallica L.   English Catchfly.   Native?
Cultivated land, roadsides and waste places.   Abundant.
M.A.T.B.MN. : -.-.-.-.-.   First record: Bree, 1831.
This is one of the abundant and characteristic weeds of the bulbfields, and prior to the introduction of the bulb industry cornfields were the main habitat. It is occasionally found away from man-made habitats as, for example, a diminutive state which grows on the hill behind Hugh Town church. The species is very variable and is represented in Scilly by the following varieties (see Lousley in *Rep. B.E.C.*, **11**, 395–6, 1937) which are believed to have a genetic basis.

var. *gallica* (var. *sylvestris* (Schott) Aschers. and Graebn.). Flowers 12–13 mm in diameter, petals at least 5 x 5 mm, pale pink or rose, almost entire. Rare. Scilly: Salmon, C. E., *Fl. Surrey*, 173, 1931. St Mary's: Bishop in *Rep. B.E.C.*, **12**, 190, 1942; near Golf Links; Holy Vale (40); near Halangy Point (41); Star Castle Garden (408). Tresco: near Dolphin's Town. St Martin's: near Middle Town.

var. *anglica* (L.) Mert. and Koch. Flowers small, about 8 mm in diameter; petals dingy white or yellowish. Abundant in all the inhabited islands.

var. *quinquevulnera* (L.) Boiss. Similar to the last but petals with a deep crimson spot near the base. Very rare. St Mary's: Newford, 1954, Mrs O. Moyse. Tresco: bulbfield weed, Northward, 1957 (773).

The populations in Scilly thus exhibit a similar range of variation to those in Jersey.

S. dioica (L.) Clairv. (*Lychnis dioica* L.; *Melandrium rubrum* (Weigel) Garcke)   Red Campion.   Native.
Roadsides, hedgebanks, rough cliff slopes, etc.   Common.
M.A.T.B.MN. : S.-.H.TN.E.   First record: Millett, 1852.
Plants with pale pink or white flowers also occur.

S. alba (Mill.) E. H. L. Krause (*Lychnis alba* Mill.; *Melandrium album* (Mill.) Garcke   White Campion.   Native.
Roadsides and field borders.   Rare.
-.A.T.B.MN. : -.-.-.-.-.   First record: Smith, 1909.
St Agnes: rare, Smith, 1909; near Lighthouse; near Quay, Grose; Upper Town. Tresco: cliff fields, Herb. D.-S.; near Abbey. Bryher: about The Town; Southward, Grose. St Martin's: M. Knox.
S. alba x S. dioica is recorded from Tresco—In hedgerow near Abbey Gardens and edge of bulbfield above Old Grimsby, 1968, R. Lancaster.

Lychnis L.
L. flos-cuculi L.   Ragged Robin.   Native.
Marshy meadows.   Local.
M.-.-.-.-. : -.-.-.-.-.   First record: Townsend, 1864.
St Mary's: Old Town Marsh, Townsend: Higher and Lower Moors, locally plentiful.

Agrostemma L.

A. githago L.   Corn Cockle.   Casual.

Cornfield weed.   Extinct.   M.-.-.-.-.   :   -.-.-.-.-.

First and only record: St Mary's, Townsend, 1864.

Corn Cockle, a weed of Mediterranean origin, was formerly common in Britain when the seeds were introduced and distributed with cereals. No doubt it occurred in Scilly until cornfields were superseded with bulbfields and seed corn became almost free from impurities.

Saponaria L.

S. officinalis L.   Soapwort.   Denizen.

Waste ground.   Local.

-.-.T.-.-.   :   -.-.-.-.-.   First record: Bree in Forbes, 1821.

Tresco: given for this island by Bree, 1821; near New Grimsby, Townsend, 1864; near Dial Rocks and the Duckery, Herb. D.-S.

Soapwort was formerly grown for officinal purposes, and the leaves were pulverised in water for use as a detergent. Its occurrence on Tresco alone, where there was an Abbey, is likely to be due to introduction by the monks of this very persistent plant.

Cerastium L.

C. fontanum Baumg. subsp. *triviale* (Murb.) Jalas (*C. holosteoides* Fr.)   Common Mouse-ear Chickweed.   Native.

Roadsides, cliffs, pastures, consolidated dunes, etc.   Very common.

M.A.T.B.MN. : -.-.H.TN.E.   First record: Townsend, 1864.

C. glomeratum Thuill. (*C. viscosum* auct.). Broad-leaved Mouse-ear Chickweed.   Native.

Roadsides, walltops, and cultivated land.   Common.

M.A.T.B.MN. : -.-.-.TN.-.   First record: Townsend, 1864.

This species is more dependent on disturbed soil conditions than the last, and hence is found in greatest quantity in bulbfields and other 'open' associations. On the uninhabited islands it has been found only on Tean, and here it may have persisted from the time when part of this island was cultivated.

On the mainland the white petals of this species usually about equal the sepals, but in Scilly they often exceed them by 1–2 mm, thus giving the plants an unfamiliar appearance. This large petalled form, forma *macropetalum* Druce (*Rep. B.E.C.*, **7**, 31, 1923) is persistent, for example, in bulbfields near Old Town, St Mary's (606, 880).

C. diffusum Pers. (*C. atrovirens* Bab., *C. tetrandrum* Curt.)   Sea Mouse-ear Chickweed.   Native.
Loose sandy ground, bare places on thin soil over granite on cliffs and downs, walls.   Abundant.
M.A.T.B.MN. : S.AT.H.TN.E.   First record: Townsend, 1864.
Although so abundant this species is easily overlooked on account of its early flowering and size. It flowers mainly in March and April and is usually quite dried up by mid-May, though occasional plants may be found where there is a little moisture throughout the summer. The most common form is an unbranched plant, sometimes only a centimetre tall, but in suitable habitats and seasons it is branched and much taller, such as specimens 30 cm tall which I found in the kistvaen on North Hill, Samson (484). From Annet (63), on ground enriched by the droppings of sea-birds, I have collected densely caespitose much-branched plants about 6 cm across, with numerous flowers and narrow capsules longer than the calyx (=*C. tetrandrum* forma *caespitosum* Druce in Moss, *Camb. Br. Fl.*, **3**, 53.).)

Stellaria L.
S. media (L.) Vill.   Common Chickweed.   Native.
Cultivated and waste ground, etc.   Very common.
M.A.T.B.MN. : S.AT.-.TN.E.   First record: Townsend, 1864.
A very large over-wintering form found in bulbfields in early spring much resembles *S. neglecta*, but may be distinguished from it by the pale seeds *c* 1 mm in diameter with rounded tubercles (e.g. St Agnes, Cove Vean (611) and St Martin's (582).)
On the uninhabited islands chickweed is probably an introduction. On Annet I have found it only near the landing place (54), where it probably arrived as seed on the shoes of a visitor (*J. Bot.*, **77**, 198, 1939). On the Eastern Isles I have seen it only on the summit of Great Innisvouls on ground rich in guano, where it may have been introduced by a bird. Samson and Tean were formerly inhabited, and are now much visited.

S. pallida (Dumort.) Piré (*S. apetala* auct., *S. boreana* Jord.).
Lesser Chickweed.   Native.
Sand dunes.   Uncommon.

-.A.T.B.MN. : S.-.H.TN.-.   First record: Lousley, 1967.
St Agnes: M. Jaques, 1957. Tresco: M. Jaques, 1957. Bryher: Rushy Bay, 1953 (555). St Martin's: sandy shore, Lower Town, 1953 (579). Samson: 1954. Northwethel, 1957. Tean: East Porth, 1953 (620). No doubt more frequent than these records indicate but it flowers in March and April and disintegrates before most botanists arrive.

S. neglecta Weihe   Hedge Chickweed.   Native.
Damp roadside and under trees.   Rare.
M.-.T.-.-. : -.-.-.-.-. First record: Townsend, 1864—without locality.
St Mary's: near Porthloo, Grose; roadside near Tremelethen (55); Watermill Lane. Tresco: lower part of Abbey Grounds (52).

S. holostea L.   Greater Stitchwort.   Native, probably extinct.
M.-.-.-.-. : -.-.-.-.-.   First record: Millett, 1852, without locality.
St Mary's: Ralfs' *Flora*, 1879. Although Ralfs failed to include the usual mark in his Flora to indicate that he had seen the plant himself, he mentions it in a letter to Townsend in 1876.

S. graminea L.   Lesser Stitchwort.   Native.
In hedges.   Very rare.
M.-.T.-.-. : -.-.-.-.-.   First record: Ralfs' *Flora*, 1879.
St Mary's: Ralfs, 1879; in hedge on left of track going east from Bar Point, 1952, Grose (spec.); Garrison Hill, 1954, Howitt. Tresco: Back Lane, 1968, D. A. Cadbury.
This species and the last are common in Cornwall but their rarity in Scilly is paralleled in the smaller Channel Isles, though not in the larger island of Jersey.

S. alsine Grimm (*S. uliginosa* Murr.)   Bog Stitchwort.   Native.
Marshes, and the sides of pools and ditches.   Locally common.
M.-.-.-.-. : -.-.-.-.-.   First record: Townsend, 1864.
St Mary's: Holy Vale, Grose; Higher and Lower Moors; roadside near Newford; Pungies Lane (291).

Moenchia Ehrh.
    M.erecta (L.) Gaertn., Mey., and Scherb. This is given as 'common' in Ralfs' *Flora*, 1879, but no other botanist has recorded it and the record must be regarded as an error.

Sagina L.

S. apetala Ard.   Annual Pearlwort.   Native.
Pavements and the base of walls.   Rare.
M.-.T.-.-. : -.-.-.-.-.   First record: Townsend, 1864.
St Mary's: in pavement near Hugh House, Garrison Walk (406);
wall at Old Town; Hugh Town, 1852, J. Woods, Herb. S.L.B.I.
Tresco: bulbfield, Pool Lane (887).
Townsend entered this in his manuscript in 1862 as 'not common?'
but altered this to 'common' in the printed account. It may be
less rare than the records indicate.

S. ciliata Fr.   Fringed Pearlwort.   Native.
Dry bare ground on tracks, roadsides, bulbfields, walls and cliffs.
Very common.
M.-.T.B.MN. : S.-.H.-.-.   First record: Townsend, 1864.
Townsend's statement that 'In the Scilly plants the upper part of
the pedicels, and the calyx, are always glandular-hairy . . .' is too
sweeping, though such plants occur (e.g. bulbfield near The Town,
Bryher (345)). This species grows in much drier places than the
last. Further work may show that they should be treated as forms
of the same species.

S. maritima Don.   Sea Pearlwort.   Native.
Bare places on cliffs and crevices in rocks and walls.   Very common.
M.A.T.B.MN. : S.AT.H.TN.E.   First record: Townsend, 1864.
S. maritima is most common near the sea, but it also occurs inland
on walls and in bulbfields to which salt spray may be driven by the
winter storms. It is very variable in size and habit. A minute state
less than 1 cm tall occurs on the cliffs of Samson and other islands,
and var. *debilis* (Jord.) is recorded from St Martin's by Townsend,
1864, and var. *densa* (Jord.) from St Mary's, 1877 in Herb. Tellam.
These are probably states of no taxonomic value.

S. procumbens L.   Mossy Pearlwort.   Native.
On tracksides, cliffs and field borders; often in turf and usually in
wetter places than the preceding species.   Very common.
M.A.T.B.MN. : S.-.H.-.E.   First record: Townsend, 1864.
Noticed on the uninhabited islands only on Samson, St Helen's,
and Great Ganilly, on all of which it is rather scarce.

Honkenya Ehrh.
H. peploides (L.) Ehrh.   Sea Purslane.   Native.
Sand and shingle beaches, a little above normal spring tides.
Frequent.
M.A.T.B.MN. : -.-.H.TN.-.   First record: Townsend, 1864.

Arenaria L.
A. serpyllifolia L.   Thyme-leaved Sandwort.   Native.
Sand dunes.   Rare.
M.-.-.B.MN. : S.-.-.TN.-.   First record: Townsend, 1864.
St Mary's: Pelistry Bay. Bryher: Rushy Bay. St Martin's: M. Jaques.
Samson: M. Jaques. Tean: East Porth (708).
Townsend gave this as 'Common, but small, and not easily detected'
but it seems that he overstated the frequency.

Spergula L.
S. arvensis L.   Corn Spurrey.   Native.
Cultivated ground.   Common.
M.A.T.B.MN. : -.-.-.-.-.   First record: Millett, 1852.
The common plant is var. *arvensis*, but I have also collected var.
sativa (Boenn.) Mert. & Koch from a bulbfield on Tresco in 1938
(56). In Scilly the species occurs only on cultivated land and the
interesting dune plant frequent in the Channel Islands (var. *nana*
Linton) has not been found.

Spergularia (Pers.) J. & C. Presl
S. rubra (L.) J. & C. Presl   Sand Spurrey.   Native.
Paths, quarries, bulbfields and on a shingle bank.   Frequent.
M.-.T.-.MN. : -.-.-.-.-.   First record: Ralfs in litt. to Townsend,
1877 (but Townsend noted in his manuscript 'a specimen in Herb.
J. Woods gathered in Scilly. Aug. 15, 1852'. I have been unable to
trace the specimen.)
St Mary's: Ralfs, 1877; bulbfield, Halangy (463); gateway near
Parting Carn (467); London (61); cultivated field, Normandy (60);
trackside, Mount Todden (743); Porthellick strand, Polunin (146a).
Tresco: path in Abbey Grounds (354); small quarry in Racket
Town Lane (454). St Martin's: Ralfs, 1877.
This species seems to be increasing as it spreads into habitats made
available by human activity. All the habitats are on soils derived
from granite, but not all are markedly sandy.

S. bocconii (Scheele) Aschers. and Graebn. Native.
Very rare. On a bank of fine shingle.
M.-.-.-.-. : -.-.-.-.-. First record: Lousley, 1967.
St Mary's: Porthellick strand, 1953, Polunin (146). Mr Polunin
sent me specimens of this and of *S. rubra*, with which it was associ-
ated, but I have been unable to refind either species at Porthellick.
Abundant in the Channel Islands, and found in Cornwall and
Devon (and elsewhere) in England, there seems no reason why
S. bocconii should not be native in Scilly.

S. rupicola Lebel ex Le Jolis. Rock Spurrey. Native.
Rocks, roadsides and walls all over the islands. Abundant.
M.A.T.B.MN. : S.AT.H.TN.E. First record: Townsend, 1864.
The abundance of Rock Spurrey is one of the outstanding botanical
features of the Isles of Scilly. It is equally at home on walls in the
heart of the larger islands and on isolated sea-swept rocks such as
Rosevear and Mincarlo, and is one of the most ubiquitous species
in the flora.
A stout plant with very prominent stipules from a cultivated field
at Normandy, St Mary's (60) was recorded as *S. rupicola* var.
glabrescens Brèb. (*J. Bot.*, **78**, 154, 1940). This was subsequently
grown from seed from the herbarium sheet and found to be a form
of *S. rubra*.

S. marina (L.) Griseb. Lesser Sea Spurrey. Native.
By brackish ponds and ditches near the sea. Rare.
M.-.-.B.-. : -.-.-.-.-. First record: Townsend, 1864.
St Mary's: Old Town Marsh, Townsend, 1864 and Ralfs, 1879;
round the pond in the Higher Marsh, Ralfs, 1877; Old Town
Marsh, 1953, Polunin. Bryher: on mud by the Pool (59 & 334).
A specimen of this species in Herb. Townsend labelled *Lepigonum
neglectum* Kindb. gathered in Old Town Marsh by Townsend in
June 1862 has large capsules on long peduncles containing both
winged and apterous seeds. Elsewhere hybridism with *S. media*
might be suspected but that species is unknown in Scilly. *S. marina*
was probably more plentiful in Higher and Lower Moors when
these were more brackish than they are now.

Polycarpon L.
P. tetraphyllum (L.) L. Four-leaved All-seed. Native.
Roadside walls, sand-dunes and bulbfields. Common. (Fig 8,

p 155).　　M.A.T.B.MN. : -.-.-.-.-.　　First record: Miss Anne Dorrien-Smith in Thurston, 1929.

This is now common and sometimes abundant in St Mary's, Tresco, Bryher and St Martin's, and less so on St Agnes. The earliest evidence of its occurrence is a specimen from Tresco Gardens collected in April 1928 (Herb. D.-S.), and it was recorded by Miss E. S. Todd from St Martin's in 1933 and by me from St Mary's in 1936, from Bryher in 1938, and from St Agnes, the Gugh in 1939. My notes show that by 1940 it was as common as it is now. *P. tetraphyllum* has proved an efficient coloniser in many parts of the world but it seems unlikely that it could have spread to, and increased on, five islands in such a short time after its first discovery. In a few places, such as dunes near Lower Town, St Martin's it grows in natural habitats where it may have been overlooked by earlier visitors, and from these, bulbfields would provide suitable conditions for rapid spread. From bulbfields it would be an easy step to roadside walls. It is accepted as native on the mainland of Cornwall, and is abundant in all the Channel Islands (see also Lousley, 1939, p 198).

ILLECEBRACEAE

Corrigiola L.
C. litoralis L. Strapwort. Tresco: on sandy shore, April 1929, R. Meinertzhagen, Herb. Mus. Brit. The habitat and date are unlikely and the specimen probably did not come from Scilly.

Scleranthus L.
S. annuus L.　　Annual Knawel.　　Native.
Probably cultivated ground.　　Very rare.
M.-.-.-.-. : -.-.-.-.-.　　First record: Montgomery, 1854—without locality.
St Mary's: Ralfs, *Flora*, 1879. No further records.

PORTULACACEAE

Montia L.
M. fontana L.　　Blinks.　　Native.
Roadsides, by ponds and springs, and also early in the year in bulbfields even on well drained sand.　　Common.
First record: Townsend, 1864.
I am grateful to Dr S. M. Walters for determining the exsiccatae cited below.

subsp. *amporitana* Sennen (subsp. *intermedia* (Beeby) Walters;
M. lusitanica Samp.)   M.A.T.B.MN. : S.-.-.-.-.
This subspecies, of western distribution in Britain, is the common
one in Scilly:—St Mary's: Newford (137); roadside, Tremelethen
(72); lower part of Watermills Lane (410). St Agnes: Wingletang
Downs (608); The Gugh. Tresco: abundant in bulbfields below
Monument Hill. Bryher: common; by pond, Pool (561). St Martin's:
bulbfields below Lower Town (580) and Middle Town. Samson:
spring near ruins (70); by pool, south-east corner (71).
subsp. *variabilis* Walters   M.-.-.B.-. : -.-.-.-.-.
This subsp. is scattered through Britain, but is most common in
northern England and Wales. In Scilly it is apparently rare, and
in both the localities where it has been found it grows in dry
situations, on shallow soil over granite on the slopes of hillocks
and associated with *Poa infirma:*—St Mary's: Carn Friars (587).
Bryher: Pool (563).

M. perfoliata (Willd.) Howell (*Claytonia perfoliata* Donn ex Willd.)
Perfoliate Claytonia.   Established alien.
Roadsides, bulbfields and waste ground.   Abundant and a
serious pest.   M.A.T.B.MN. : -.-.-.-.-.
First record: Miss Anne Dorrien-Smith in Thurston & Vigurs, 1928.
St Mary's: Watermills, A. Dorrien-Smith, as above; abundant
at Watermills, 1939 (88); bulbfield Halangy, 1940; bulbfields,
Toll's Porth, 1947, G. Grigson; waste ground Hugh Town, 1952,
A. Conolly; Porth Mellon and wall near Porthloo, 1956, J. Matheson; on wall, Old Town, 1953; abundant in bulbfields throughout
the island, 1954. St Agnes: High Town and near Lighthouse, 1953.
Tresco: field border between Church and Abbey, believed eradicated,
1940, A. A. Dorrien-Smith. Bryher: Rushy Bay, 1953. St Martin's:
abundant in bulbfields below the School, 1953 (575); in the greatest
profusion in bulbfields all along the south coast from Lower Town
to the east end of Higher Town Bay, 1953.
A native of the west coast of North America, this species has been
established in Britain for over a century and spreads rapidly on
sandy soils. It is sometimes grown as an early salad plant and may
have been introduced into Scilly for this purpose. From 1928 its
spread in Scilly was spectacular, but it is only conspicuous in early
spring and summer visitors may fail to see it.

Portulaca L.

P. oleracea L.   Pot Purslane.   Alien.

Weed of arable land.   Rare.   M.-.T.-.-. : -.-.-.-.-.

First record: Anon—'Naturalised at Tresco, Scilly Isles', *Rep. B.E.C.*, **7**, 868, 1926 and 'Scilly Isles, sea shore, Aug. 5, 1925, G. C. Druce' in Herb. Kew. (Druce is not known to have visited Scilly). St Mary's: abundant in bulbfield near Parting Carn, 1957, Miss D. M. Turner. Tresco: see above; abundant in Abbey Gardens as a weed; abundant in field near Pool Road, 1970.

Pot Purslane is a native of the south-western parts of the United States, was formerly much grown as a salad and pot-herb, and is now a cosmopolitan weed. It is established in Jersey, and commonly so in western France, and it thrives on warm sandy soils. Major A. A. Dorrien-Smith said it flourished on ground which had been manured with seaweed, and Mr B. Watts, the tenant of Parting Carn, thought it had been introduced to his farm in seaweed used as a top dressing for the bulbs.

Although some of the plants in Scilly are erect they all appear to be subsp. *oleracea*, and subsp. *sativa* (Haw.) Celak., which has larger obovate leaves and is very robust, does not occur.

AIZOACEAE

All the species belonging to this family which occur wild in Scilly were until recently included in the large genus *Mesembryanthemum*, which has now been divided into many small genera. These succulents ('mesems' as they are commonly called in Scilly) are nearly all natives of South Africa and very sensitive to frosts. A few species are hardy in the Channel Islands and south-west England but Scilly is the only part of Britain where a wide selection of this genus can be grown in the open throughout the year without protection.

The cultivation of these colourful plants in Tresco Abbey gardens goes back a long way. Augustus Smith was receiving cuttings from Kew in 1849 and even then had a collection of mesems which justified inviting Sir William Hooker to Tresco. By 1852 there were 50 species at Tresco Abbey, which had grown to 96 listed by A. Henwood Teague in 1889, and 120 by the end of the century. A manuscript garden list was started in 1935 and twelve years later this included 153 species (Hunkin, 1947). A sketch made in 1862 shows that cushions of these plants were growing on the hillside beyond the garden, a forerunner of the widespread planting which has taken place since. Since they can be propagated from small pieces broken off, it is not surprising that the more hardy species are commonly grown in cottage gardens on all the inhabited islands. They also occur on walls in the lanes and on the cliffs where they persist in some cases in competition with native vegetation. Of the large number of species to be seen, only those which seem to be successfully established are included in the following account.

Mesembryanthemums are exceedingly difficult to name. They make poor herbarium specimens and are poorly represented in the national collections. They are still inadequately studied even in South Africa and new species are constantly being described. Many of the Tresco plants have been imported direct and do not appear in horticultural works, and none of the accounts cover them all. To assist the reader to recognise the species in the field I have added brief descriptions, but much more work needs to be done on these plants which add so much colour to the scenery in Scilly.

Carpobrotus N. E. Brown
C. edulis (L.) N. E. Br.   Hottentot's Fig.   Naturalised alien.
Cliffs and dunes.   Very common.   M.A.T.B.MN. : -.-.H.-.-
First record: Davey, 1909 (as *Mesembryanthemum aequilaterale* Haw.), but see below.
St Mary's: common in the vicinity of Hugh Town; Porth Mellon; Porthloo; Garrison; Old Town; Porth Minick; Watermill Cove, etc. St Agnes: Gugh Bar by Gugh Farm; Porth Conger. Tresco: Appletree Banks; about Abbey Farm; Plumb Island; Old Grimsby; New Grimsby, etc. Bryher: in several places on the shores. St Martin's: near Yellow Rock; Little Bay; Great Bay. St Helen's: slope above monastery; abundant all over the island. Round Island: abundant.
This is the common 'ice plant' with large usually yellow flowers which turn pink as they go over, and thick woody trailing stems with fleshy leaves, triangular in section and three inches or more long. It is now thoroughly established on the cliffs of south-west England and in the Channel Islands, and even more so in Scilly. Grown at Tresco Abbey for many years, it was planted by Augustus Smith in the middle of last century with Marram Grass to stabilise the sand-dunes (*Scillonian*, **9**, 21 seq., 1935). By 1921 it was 'common on most of the islands', and in 1924 Miss E. Farthing reported it from St Mary's, Tresco, St Martin's and St Agnes. No doubt the

initial wide distribution was due to planting, as, for example, when the lighthouse keepers planted it on Round Island, but further colonies have been started from bits of the plant transported by gulls. These are probably taken for the construction of their rough nests and start plants on the tops of carns and other places where humans would be unlikely to introduce it.

Once introduced in a suitable habitat Hottentot's Fig increases vegetatively very rapidly, competes successfully with most native plants, and covers a large area in a remarkably short time. This is especially the case after fire. Grigson reported a large patch on the slope above the monastery on St Helen's in 1944 on ground where the vegetation had been burnt off by German incendiary bombs in 1941. Three years later it had greatly increased (Grigson, 1947) and by 1957 it was abundant in many widely scattered colonies on the burnt ground and also on rocks over much of the island. Periods of drought also facilitate rapid increase since they reduce the height of bracken and other competitors. *Carpobrotus edulis* is seriously cut back by severe frosts on the rare occasions they occur in Scilly. In such years, as for example after the winter of 1942/43, long ropes of dead stems and leaves are to be seen in exposed places, but there are always parts of the plants left alive and these make good most of the setback during the following summer.

C. edulis with yellow flowers presents no identification problems but it is sometimes found with purple blossoms and care is necessary to distinguish these from several closely allied species which are discussed below. *C. edulis* has the flowers *c* 8 cm across, the perianth tube is club-shaped, and the leaves are green, 5–13 cm long and very much the same width throughout almost the whole of their length. It seems likely that the purple flower colour has been introduced into early British populations by hybridisation with a species which normally has flowers of that colour such as *C. aequilaterus* or *C. acinaciformis*.

Carpobrotus is a widely distributed genus, being found in South Africa, Australia, New Zealand, Chile and California, and a careful revision by someone with access to sufficient living material is very much needed. In the meanwhile, as the authority N. E. Brown remarked when listing the species which have been described, 'it is scarcely possible to identify cultivated plants from the descriptions given or to understand how they differ' (*J. Bot.*, **66**, 323, 1928). The following account must be regarded as tentative.

C. aequilaterus (Haw.) N. E. Br. (*Mesembrythemum aequilaterum* Haw.; *M. aequilaterale* Haw.). Material from Cornwall and Scilly now referred to *C. edulis* was first identified as this Australian species in error. It has smaller (*c* 6 cm across) purplish-red flowers, an obconical perianth tube about 2 cm long, and the flowers are subtended by bracts. Plants which may be this grow by Carn Near, Tresco (791).

C. acinaciformis (L.) L. Bolus. (Fig 10.) This native of the Cape of Good Hope, where it grows on sandy flats by the sea, has been known in England since 1732. Plants sent from Tresco by Augustus Smith were figured as this in *Curtis's Botanical Magazine* in 1865 (*t.* 5539). It is to be separated from *C. edulis* by the shorter sub-globose perianth tube, pale glaucous leaves which are scimitar shaped (i.e. broadest above the middle), and the bright magenta flowers. It grows on the dunes of St Martin's (e.g. Laurence's Bay, 869), and Appletree Banks, Tresco, and at Porthloo, St Mary's.

C. deliciosus (L. Bolus) L. Bolus (in *Ann. Bolus Herb.*, **4**, 41, 1926). It has been suggested by Batten and Bokelmann in 'Wild Flowers of the Eastern Cape Province', 1966 that this species, which is common as a native on dunes in the Eastern Cape Province and has been in cultivation for two centuries, is the one known as 'Sally-my-handsome' in England and used for stabilising dunes in Britain as well as Australia and Portugal. I have compared material from the type locality (Riversdale, Herb. Kew) with specimens sent in 1895 from Tresco Abbey for a drawing in *Curtis's Botanical Magazine* (*t.* 8783) and these agree in the leaves and flowers, but the fruits, on which the species is based, are not available. Ripe fruits occur on Hottentot's Figs in Scilly but so far I have been unable to find any with the peduncles constricted at the apex and altering their position from being erect in flower, to horizontal, then curved downwards to finally curving upwards from about the middle when the fruit is ripe. Further study may show that this, and several additional species of *Carpobrotus* which have been grown at Tresco Abbey, are to be found wild in Scilly.

Drosanthemum Schwant.
D. floribundum (Haw.) Schwant.   Naturalised alien.
Coastal rocks and banks; inland on walls.
Local. (Fig 10, p 217)   M.-.T.-.-. : -.-.-.-.-.
First record: Lousley, 1967.
St Mary's: well established below Middle Carn, Old Town and on coastal rocks near Woolpack Point, Garrison (682); walls and banks, Porthloo (741); walls near Normandy. Tresco: common and thoroughly established; New Grimsby (628); Old Grimsby; etc.

This is a sprawling plant with woody stems and pale mauve flowers recalling those of the smaller Michaelmas Daisies. The leaves are greyish, about ½ in long, covered with white papillae, and the upper stems have white, reflexed bristle-like papillae. A native of Cape Province, it has long been grown in Scilly. There is a specimen dated 1875 from the Old Churchyard, St Mary's where it was presumably cultivated (Herb. Mus. Brit.). It is now thoroughly established in St Mary's and Tresco, but on St Agnes, Bryher and St Martin's I have noticed it only near houses.

Oscularia Schwant.
O. deltoides (Mill.) Schwant.   Established alien.
Walls.   Common. (Fig 10, p. 217)
M.-.T.-.MN. : -.-.-.-.-.   First record: Lousley, 1967.
St Mary's: common on walls about Hugh Town, Old Town, Parting Carn, etc. Tresco: sea-wall, New Grimsby (627); Borough Road (792); near Appletree Carn; walls of Abbey Gardens, etc. St Martin's: Lower Town, etc.
This plant has deltoid, keeled, pruinose leaves and small purple flowers which appear in April. Although it is found on walls far from houses, maintaining itself without assistance, I have no evidence that it spreads by natural means. An allied species, *O. caulescens* (Mill.) Schwant. is also a native of Cape Province, and distinguished from *O. deltoides* by the larger leaves being usually toothed at the sides, and not on the keel. Material of *O. caulescens* sent from Tresco Abbey Gardens in 1895 was compared by N. E. Brown with the type in the Dillenian herbarium. It is stated to be hardy at Plymouth. The two species are not easy to distinguish when dried, and my material has some of the characters given for *O. caulescens*, but is not that species as now grown in the Abbey Gardens.

Aptenia N. E. Br.
A. cordifolia (L.f.) N. E. Br. (*M. cordifolium* L.f.)   Established alien.
Walls.   Rare.
M.-.-.-.MN. : -.-.-.-.-.   First record: Lousley, 1967.
St Mary's: wall in lane above Porthloo (739, 828); wall inside field between Hugh Town and Old Town (793). St Martin's: abundant on wall at High Town.

A. cordifolia is a well known pot-plant, with sprawling papillose stems and opposite, green, fleshy flat subcordate leaves bearing papillae. The flowers are small (about ½ inch across), purplish-red and appear in June–August. It is a native of Cape Province which is naturalised in Madeira, St Helena and Australia. In Scilly it persists and spreads on walls away from houses but there is no evidence that it is distributed by natural means.

Lampranthus N. E. Br.

L. falciformis (Haw.) N. E. Br. (*M. falciforme* Haw.)   Established alien.

Walls.   Common.

M.-.T.-.MN. : -.-.-.-.   First record: Lousley, 1967.

St Mary's: common on walls; Old Town Lane (681); Porthloo (740); etc. Tresco: very common on walls and hedges, and often far from houses; Middle Down (790); etc. St Martin's: walls.

A native of Cape Colony, this species may be recognised by its large (1½ inch across) brilliant mauve flowers in July and August, and grey-green leaves about ¾ inch long with translucent dots. There are specimens in Herb. Kew sent from Tresco Abbey Gardens in 1895 and it has been found to be hardy at Plymouth and Bournemouth. It was collected on Lundy Island as naturalised in 1908 and 1911 (Herb. Kew).

Several other species of *Lampranthus* grown in Tresco Gardens, including *L. conspicuus* (Haw.) N. E. Br. with brilliant scarlet-red flowers, and *L. spectabilis* (Haw.) N. E. Br. with bright mauve purple flowers, are planted on walls and may persist.

Ruschia Schwant.

R. caroli (L. Bol.) Schwant.   Established alien.

Walls.   Local.

M.-.T.-.-. : -.-.-.-.   First record: this *Flora*.

St Mary's: walls about Old Town, 1959 (829); abundant round Old Town, 1967 (867); roadside wall, Holy Vale, 1963 (835); roadside wall near Tremelethen competing with brambles, 1967; Carn Friars, 1967. Tresco: seen in several places on walls.

This is a woody, bushy plant about 2 ft tall, with punctate, sub-terete leaves about 1 inch long turning purple, and flowers about 1 inch across, 3–4 terminal on branches, with purple petals. There is a specimen sent in 1897 as *sarmentosum* from Tresco Abbey in Herb. Kew. It flowers in April–May.

Disphyma N. E. Br.

D. crassifolium (L.) L. Bolus (*M. crassifolium* L.)   Established alien.   On sea-walls. Rare (Fig 10, p 217)

M.-.T.-.-. : -.-.-.-.-.   First record: Lousley, 1967.

St Mary's: by sea, Old Town (626). Tresco: on the sea-wall at New Grimsby (765 & 824).

When I first found this plant on the sea-wall at New Grimsby in 1936 elderly inhabitants told me that it had been there as long as any of them could remember. There was then a very large patch, draping down on to the beach below, and this has maintained its ground ever since. The leaves are fascicled, very fleshy, round in section and reddish, on long pale stems rooting at the nodes. The flowers are sessile, solitary and bright purple. It is a native of South Africa where, as in Scilly, it occurs near the sea.

Erepsia N. E. Br.

E. heteropetala (Haw.) Schwant. (*M. heteropetala* Haw.) Established alien.

On rocks.   Rare. (Fig 10, p 217.)

M.-.-.-.-. : -.-.-.-.-.   First record: Lousley, 1967.

St Mary's: Buzza Hill Quarry (794).

This species is abundant on the rock-face of the long disused quarry at Buzza Hill where it spreads freely from seed. The leaves are about 1 inch long, sabot-shaped and glaucous, with the keel and margins strongly serrate. The flowers, which are to be seen in June, are small with the petals hardly exceeding the calyx. This species, which is a native of South Africa, has been found to be hardy at Plymouth.

AMARANTHACEAE

Amaranthus L.

A. retroflexus L.   Pigweed.   Established alien.

Cultivated fields, roadside and sand-dune.   Local.

-.-.T.-.-. : -.-.-.-.-.   First record: Lousley, 1967.

Tresco: abundant in fields near Pool Road, 1952 (507), 1970; Borough, 1953, Polunin; Appletree Banks, 1953, Polunin; roadside between Old and New Grimsby, 1956 (717).

This alien of North American origin is now an established weed in Tresco, but has not yet spread to other islands.

A. hybridus L.   Casual in arable field.
-.-.T.-.-. : -.-.-.-.-.   First record: Lousley, 1967.
Tresco: in small quantity growing with *A. retroflexus* in field near
Pool Road, 1952 (506).

CHENOPODIACEAE

Chenopodium L.
  C. bonus-henricus L.   Good King Henry.   Denizen.
By farm buildings.   Very rare.
-.-.T.-.-. : -.-.-.-.-.   First record: Lousley, 1967.
Tresco: By farm buildings, 1960, Miss R. J. Murphy.

C. polyspermum L.   Many-seeded Goosefoot.   Casual?
Arable field.   Very rare.
-.-.T.-.-. : -.-.-.-.-.   First record: Lousley, 1967.
Tresco: weed in cultivated field by Pool Road, 1952. This species,
which is rather rare in Cornwall, appears to be a recent introduction
in Scilly.
C. vulvaria L. Reported by Ralfs in litt. to Townsend, 1876 as 'C. *olidum*—
all the sandy shores'. This is an obvious error for forms of *Atriplex
laciniata* and no doubt Ralfs' record for St Mary's in his *Flora* was
also a mistake.

C. album L.   Fat Hen.   Native.
Cultivated ground and waste places.   Common.
M.A.T.B.MN. : -.-.-.-.-.   First record: Townsend, 1864.
C. opulifolium Schrad. ex Koch and Ziz. Error. This is recorded for
Scilly by Miss E. S. Todd in *Rep. B.E.C.*, **10**, 539, 1934 and there is a
specimen in her herbarium at Swindon so labelled from 'Farmyard,
St Martin's, August 1933'. This was named by W. H. Pearsall but is an
ordinary form of *C. murale*.

C. murale L.   Nettle-leaved Goosefoot.   Native.
Roadsides, waste places, sand-dunes and cultivated land. Common.
M.A.T.B.MN. : -.-.-.-.-.   First record: Townsend, 1864.
In Scilly this species grows throughout the year and plants in
flower and fruit are commonly found in spring and early summer.
The life-cycle is thus different from the one usual in England,
where it germinates in spring, flowers from July onwards, and
fruits until cut back by frost.

C. rubrum L.   Red Goosefoot.   Native.
Sandy margins of pools, cultivated ground, and on a rocky shore.
Local. (Map 6, p 150.)

M.-.T.B.-. : S.-.-.-.-.   First record: Ralfs' *Flora*, 1879.

St Mary's: Old Town Marsh, Polunin. Tresco: Herb. D.-S.; Old Grimsby, Ralfs, 1879; Abbey Gardens; plentiful on sandy mud by Abbey Pool. Bryher: margin of Great Pool. Samson: scattered along the eastern shore, and on damp ground in southeast corner.

C. botryodes Sm. Error. Vyvyan, 1953, 160 lists '*C. rubrum botryodes*' on the evidence of a specimen in Herb. D.-S. determined by Edinburgh. The specimen is *C. rubrum*.

Beta L.

B. maritima L. subsp. *maritima* (L.) Thell.   Sea Beet.   Native.

Rocky coasts.   Abundant.

M.A.T.B.MN. : S.AT.H.TN.E.   First record: Millett, 1852.

Sea Beet is especially abundant on the rocks of the smaller islets,

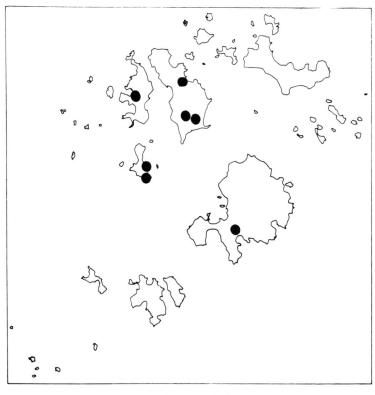

Chenopodium rubrum, Red Goosefoot

such as Rosevear and Mincarlo where it endures constant drenching with salt-spray when the sea is calm, and frequent douches of salt-water during storms.

B. trigyna Waldst. & Kit. Recorded in Vyvyan, 1953, 160 on the evidence of a specimen in Herb. D.-S. determined by Edinburgh. This specimen is unlocalised.

Atriplex L.

A. patula L.   Common Orache.   Native.
Cultivated land, roadsides, waste places and shores.   Common
M.A.T.-.MN. : -.-.-.-.-.   First record: Ralfs' *Flora*, 1879 (as '*A. angustifolia*').

A. hastata L. (*A. deltoidea* Bab.).   Halbert-leaved Orache. Native.
Shores, waste places, and cultivated land.   Very common.
M.A.T.B.MN. : -.AT.-.TN.-.   First record: Townsend, 1864, as *A. deltoidea* Bab.
This is the commonest species on many beaches.

A. glabriuscula Edmondst. (*A. babingtonii* Woods).   Babington's Orache.   Native.
Shingle and sandy shores.   Common.
M.A.T.N.MN. : -.AT.-.-.-.   First record: Townsend, 1864 as *A. babingtonii* Woods.
The mixed populations of these three species on the shores of Scilly are extremely puzzling. Intermediates abound and, while some of these may be hybrids, the problem is greatly complicated by the polymorphism of the parents.

A. laciniata L. (*A. sabulosa* Rouy; *A. arenaria* Woods).   Frosted Orache.   Native.   Sandy shores just above high-water mark. Very common.
M.A.T.B.MN. : -.AT.-.TN.E.   First record: Townsend, 1864 (but there is a specimen collected from Hugh Town by J. Woods in 1852 in Herb. S.L.B.I.).
The abundance of this species is one of the botanical features of the beaches of Scilly, and it also occurs occasionally inland on heaps of sea-weed for use as top-dressing in the bulbfields. There is considerable variation in the size and shape of the leaves and whether they are sharply toothed or subentire. Seedlings appear in late May and early June.

A. hortensis L. Garden Orache. Major A. A. Dorrein-Smith drew my attention to this in Tresco Abbey Gardens and assured me that although they occasionally use it as spinach, it had never been deliberately sown or planted to his knowledge, but maintained itself from seed as a weed.

Salsola L.
S. kali L.   Prickly Saltwort.   Native.
Sand and shingle beaches.   Frequent.
M.A.T.B.MN. : S.AT.-.TN.-.   First record: Townsend, 1864.
St Mary's: Porthellick; Bar Point. St Agnes: Dropnose Porth. Tresco: Pentle Bay, Grose; Appletree Bay, Polunin. Bryher: Southward Bay; cove on south-west coast; beach below Church. St Martin's: Great Bay; Higher Town Bay; White Island. Samson: rare. Annet. Tean: 1959, Univ. Ldn. Explor. Soc.

MALVACEAE

Malva L.
M. moschata L.   Musk Mallow.   Casual.
Laneside.   Very rare.
M.-.-.-.-. : -.-.-.-.-.   First record: Dallas in Lousley, 1940.
St Mary's: High Cross Lane, a single plant, Dallas. Although rather common in Cornwall, the rarity of this species in Scilly is comparable to the Channel Islands where it is also erratic in appearance.

M. sylvestris L.   Common Mallow.   Native.
Waste ground, roadsides and about buildings.   Common.
M.A.T.B.MN. : -.-.-.TN.-.   First record: Millett, 1862.
The occurrence on Tean may be a relic of the time when the island was inhabited.

M. neglecta Wallr. (*M. rotundifolia* auct.).   Dwarf Mallow. Native.
Waste places, especially about buildings.   Common.
M.A.T.B.MN. : -.-.-.-.E.   First record: Townsend, 1864.
Recorded from Great Ganilly, Eastern Isles, by Dallas.

M. pusilla Sm.   Casual.   Sandy roadside.
Very rare.   -.-.-.-.MN. : -.-.-.-.-.   First record: Lousley, 1940.
St Martin's: near New Quay, 1936 and 1939, not seen recently.

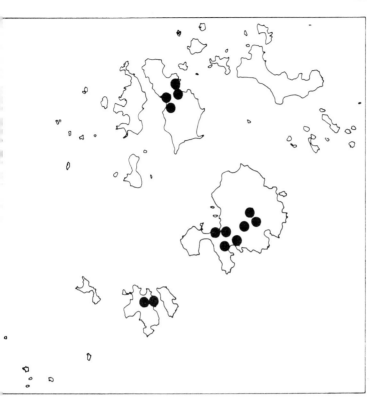

Lavatera cretica, Cretan Mallow

Lavatera L.

L. arborea L.   Tree Mallow.   Native.

Maritime rocks and cliffs.   Common.

M.A.T.B.MN. : -.AT.H.-.-.   First record: Townsend, 1864.

Tree Mallow is most at home on rough, sea-swept granite rocks, where it is sometimes the main constituent of the flora. Thus it covers the upper parts of Melledgen and Rosevear, is abundant on Mincarlo, and in quantity on Annet. Grigson states that the stems are much used by shags on Rosevear in the construction of their nests.

L. cretica L. (*L. sylvestris* Brot.).   Cretan Mallow.   Native.

Waste ground, old quarries, roadsides and bulbfields.   Locally common. (Fig 6, p 95; Map 7, p 153).

M.A.T.-.-. : -.-.-.-. First record: Curnow in *Bot. Loc. Rec* *Club Rep.* **1876,** 159, 1877.

This species was discovered by Curnow, Ralfs and Tellam in July 1873 and, after several brief reports, a full account by Trimen appeared in 1877 (*J. Bot.*, **15,** 16, 56 & 257). It was then plentiful on Tresco, St Agnes (200–300 plants at Higher Town, Ralfs), and on St Mary's 'Quite common around Hugh Town both toward the church and Old Town Bay' (Herb. Mus. Brit.). In St Mary's it has extended its ground a little but in Tresco and St Agnes the known localities are very much those reported 90 years ago. It is significant that it has not spread to the other inhabited islands and although it is restricted to disturbed ground, the history is that of a native rather than of an introduced annual. *L. cretica* is a native of the Mediterranean region and Near East, extending up the west coast of France to the Channel Isles and West Cornwall in habitat similar to those in which it occurs in Scilly.

Its appearance is very dependent on seasonal climatic condition and while I have most usually found it in quantity in April, May and June, or in early autumn, it can also be seen in other months. In my London garden it germinates in warm wet springs over a fairly wide range of dates, and in warm summers it is able to produce fruit in less than three months. In Scilly it most commonly germinates in autumn and winter, commences to flower in April (April 8, 1953 is my earliest date) and dies off rapidly at the end of June or early July. This life-cycle is well adapted to the system of cultivation of the bulbfields. If the summer is a wet one a second generation may be found in flower in the autumn, or occasional plants in the summer.

Many botanists fail to find *L. cretica* owing to confusion with *Malva sylvestris,* and as most of the characters are not apparent in dried material, or readily obtained from books, the following descriptions based on fresh material in Scilly should be useful

---

Fig 8 Bulbfield Weeds: 1 *Ranunculus marginatus* var. *trachycarpus* (fruit x1½, achene x3½). 2 *Ranunculus muricatus* (achene x3). 3 *Fumaria occidentalis* (flower x2, fruit x4). 4 *Crassula decumbens.* 5 *Polycarpon tetraphyllum*

1

2

3

4

5

L. cretica has a less coarse more graceful habit, the leaves are greyer and less rugose, the hairs on the stem shorter, and the lobes of the epicalyx are obtuse and broader (*c* 6–7 mm) and joined below for about a third of their length. The petals are much paler, a pale mauve, and of thinner almost transparent texture, almost parallel-sided, about 1.8 cm long and 7 mm wide. The flowers thus measure about 3.8 cm in diameter when fully opened out, but are usually only spreading. In contrast, *Malva sylvestris* has narrow (*c* 4 mm), lanceolate, acute epicalyx lobes which are free to the base. The petals are usually purple, about 2.2 cm long, and 1.5 cm wide at their broadest part, which is near the apex. The flowers are thus much larger, measuring about 5.4 cm in diameter, and commonly open out flat.

LINACEAE

Linum L.

L. bienne Mill. (*L. angustifolium* Huds.)   Wild Flax.   Native
Lanesides, field margins and waste ground.   Local.
M.-.-.B.-. : -.-.-.-.-.   First record: Millett, 1852.
St Mary's: sandy ground near the sea east of Hugh Town, Townsend, 1864; small green near Maypole (86); common in lanes near the Telegraph; Old Town; Holy Vale; common about Pelistry; Bar Point, Grose. Bryher: Herb. D.-S.

L. catharticum L.   Fairy Flax.   Native.
Consolidated dune and a roadside.   Very rare.
M.-.-.-.MN. : -.-.-.-.-.   First record: Townsend, 1864.
St Mary's: near Tremelethen, Grose. St Martin's: to the north of the island, Townsend; fixed dune on slopes of Great Bay.

Radiola Hill

R. linoides Roth   All-seed.   Native.
Bare, damp, and usually sandy, places.   Common.
M.A.T.B.MN. : S.-.H.-.E.   First record: Townsend, 1864.
Townsend's statement that it is 'Abundant everywhere, growing irrespective of elevation or soil' is an exaggeration, but it is widespread and, in the humid atmosphere of Scilly, less restricted to places where water has stood during the winter than is usual elsewhere.

GERANIACEAE

Geranium L.

G. pyrenaicum Burmn. fil. Recorded by H. Boyden in Tonkin & Row, 1893 and repeated in Davey, 1909 but almost certainly an error.

G. dissectum L.   Cut-leaved Cranesbill.   Native.
Waste ground, roadsides, and borders of fields.   Very common.
M.A.T.B.MN. : -.-.-.-.-.   First record: Townsend, 1864.

G. molle L.   Dovesfoot Cranesbill.   Native.
Waste and cultivated ground, roadsides and sandy turf.   Common.
M.A.T.B.MN. : S.A.H.TN.E.   First record: Millett, 1852.
White-flowered plants are frequent on St Martin's (87).

G. pusillum L.   Small-flowered Cranesbill.   Native.
Sandy waste ground.   Rare.   M.-.-.-.-. : -.-.-.-.-.
First record: H. Boyden in Tonkin & Row, 1893.
St Mary's: sandy ground, 1938 (100); Porth Mellon, M. Jaques.

G. robertianum L. Herb Robert. Recorded for Scilly by Millett, 1852 and Boyden, 1893 and for St Mary's by Ralfs, 1879, but no later records. As a specimen exists of the next species, which would not have been distinguished by earlier botanists, it is likely that *G. robertianum* has not yet been found.

G. purpureum Vill.   Lesser Herb Robert.   Native.   Extinct?
First record: Baker, H. G. in *Watsonia*, **3**, 162 & 165, 1955.
Prof. Baker cites a specimen collected by Curnow in Scilly in 1878 which is now in the herbarium of the National Museum of Wales, Cardiff. In view of the close association between Curnow and Ralfs it is likely that the latter's 1879 record for *G. robertianum* was based on the same locality and the plant probably occurred in St Mary's. It has not been reported since. The distribution of *G. purpureum* ranges through south-western and western Europe to the Channel Isles, southern England, southern Ireland and Carmarthen.

Erodium L'Herit.

E. maritimum (L.) L'Herit.   Sea Storksbill.   Native.
Bare ground on shallow soil over granite on cliffs, walls, hillocks, and on tracks in Callunetum.   Abundant. (Plate p 225).
M.A.T.B.MN. : S.AT.H.-.E.   First record: Millett, 1852.

E. moschatum (L.) L'Herit.   Musk Storksbill.   Native.
Cultivated land, roadsides, waste places, etc.   Abundant.
M.A.T.B.MN. : -.-.-.-.-.   First record: Millett, 1852.

This, like the last species, occurs in Scilly in greater abundance than in any other part of the British Isles known to me. Townsend, 1864 gives it as 'occasional' and it seems likely that the introduction of the bulb industry has provided exceptionally favourable conditions for its increase.

E. cicutarium (L.) L'Herit.   Common Storksbill.   Native.
Sand-dunes and sandy cultivated ground.   Common.
M.A.T.B.MN. : S.-.H.TN.E.   First record: Townsend, 1864.
subsp. *cicutarium* (subsp. *arvale* Andreas) is frequent in the bulb-fields and under this comes a plant collected by A. Somerville and confirmed by A. Bennett as var.*chaerophyllum* (Cav.) in Herb. Mus. Brit.
subsp. *dunense* Andreas (*E. lebelii* Jord.) is common on coastal dune. To this belongs the plant collected by Townsend as *E. pilosum* Bor. from sandy ground, St Martin's. (Townsend, 1864 and Herb. Townsend—*teste* C. C. Babington.)

E. glutinosum Dumort.   Native.
Sandy beach.   Very rare.
-.-.-.-.-. : -.-.-.-.-.E.   First record: Lousley, 1967.
Eastern Isles: Great Ganilly, 1954 (711) associated with *E. cicutarium* subsp. *dunense* with which it has been confused elsewhere.

Pelargonium Burmann
P. tomentosum Jacq.   Established alien.
Edge of woodland and in bracken.   Very rare.
-.-.T.-.-. : -.-.-.-.-.   First record: E. Brown, 1938.
Tresco: 'Escaped from Abbey Gardens and is successfully fighting the native vegetation even bracken', E. Brown in *J. Roy. Hort. Soc.*, **63**, 571, fig. 166, 1938.
This shrubby plant, about a metre tall, with a scent resembling peppermint, grows on the edge of Abbey Wood just beyond the wall of the gardens. Major A. A. Dorrien-Smith assured me that it perpetuated itself from seed, and this was confirmed when I saw it again in 1963 when all the old plants had been killed off by exceptionally severe winter weather, and many seedlings were growing around them.

OXALIDACEAE

Oxalis L.
O. acetosella L.   Wood Sorrel.   Native.

Extinct? M.-.-.-.-. : -.-.-.-. First record, Millett, 1852.

The Misses Millett gave no locality. It was also recorded by Ralfs (in litt. to Townsend, 1876) and from St Mary's (Ralfs ex Marquand, 1893). The leaves of the introduced species of *Oxalis* which are now so common in Scilly are often mistaken for Wood Sorrel, but they are not known to have been brought into the islands as early as these records.

O. exilis A. Cunn. (*O. corniculata* L. var. *microphylla* Hook. f.) Naturalised alien.
Garden weed and dunes. Local.
-.-.T.-.-. : -.-.-.-. First record: Lousley, 1967.
Tresco: in short maritime turf, 1929, R. Meinertzhagen as *O. corniculata* var. *minor* det. D. P. Young, in Herb. Mus. Brit.; abundant weed in Abbey Grounds (236); plentiful on Appletree Banks on consolidated dune (378).

O. megalorrhiza Jacq. (*O. carnosa* Molina)   Naturalised alien.
On rocks and walls.   Local. (Fig 11, p 251.)
M.-.T.-.-. : -.-.-.-.
First record: D. P. Young in *Proc B.S.B.I.*, **1**, 578, 1955.
St Mary's: Porthcressa, Young 5143; and walls elsewhere in Hugh Town; thoroughly established on walls, Porthloo Lane (625); hedgebank far from houses, Pelistry Bay, J. E. Raven and Lousley (680); Tresco: rocks and walls between Old and New Grimsby, 1940 and 1952 (525); Dolphinstown; Back Lane; abundant about Dial Rocks in field hedges and walls in the lane; abundant around Tresco Abbey Gardens.
This species is a native of Chili, where it grows mainly on coastal rocks. It was in cultivation at Tresco Abbey in 1879 (as *O. crassifolius*, painting by Mrs Le Marchant at Tresco Abbey), and this is the origin of the plants now quasi-wild. No doubt some of the colonies were started by planting bits but it spreads from seed. The stem and leaves are extremely succulent and the plant is very frost-sensitive.

O. europaea Jord. Tresco: weed in Abbey Gardens, 1968, R. Lancaster. This may occur more widely but these are the only specimens I have seen.

O. tetraphylla Cav. Tresco (894), 1969, R. Lancaster. In Jersey this was first collected as 'Four-leaved Clover', for which it might well be mistaken, and it has proved a serious weed. I have no details of the habitat in Tresco.

O. articulata Savigny (*O. floribunda* Lehmann). Naturalised alien.
Bulbfields, field borders, sand-dunes and roadsides—often under bracken.   Very common. (Fig 11, p 251.)
M.-.T.B.MN. : -.-.-.-.-.
First record: Lousley, 1939 (as *O. violacea* L.).
St Mary's: common. Most abundant in bulbfields near Bar Point (where I noticed it first), Halangy, and Helvear (where it chokes some bulbfields almost to the exclusion of other plants). Tresco: Appletree Banks (377); bulbfields, Pool Road; abundant in bulbfield, Old Grimsby. Bryher: thoroughly naturalised under bracken by the rough track from The Town to Pool, 1938 (235 & 231); bulbfield, Town Bay (709); near sea midway between Pool and Samson Hill, Grose; Northward; laneside near Pool. St Martin's: in Ammophiletum, Higher Town Bay; fields below School; in Ammophiletum, Great Bay; abundant in fields below Middle Town; west of Middle Town; near Higher Town Quay.
This species is a native of east temperate South America. I first found it in 1938 on Bryher, but C. Gibson informed me that he noticed it on Middle Town sands, St Martin's about 1925 though it was a long time before it spread from the sands into cultivated ground. Since about 1945 its spread has been extremely rapid and it is now a serious pest in bulbfields. It was probably first brought into the islands in the 1890's when Trevellick Moyle says a bulb salesman sold his uncle Oxalis as a good border plant for the garden—'Better Mr Barr had thrown them in the Thames or Uncle put them on a bonfire' (*Scillonian*, **158**, 101, 1964.). There seems to be a difference between the ecological requirements of *O. articulata* and *O. pes-caprae*. So far as I am aware they seldom occur together and there are fields on St Mary's and Tresco where they both occur with a sharp line of demarcation between them.

O. pes-caprae L. (*O. cernua* Thunb.).   Mock Shamrock, Cape Sorrel, Bermuda Buttercup.   Naturalised alien.
Bulbfields, walls, hedgebanks, roadsides, etc. (Fig 11, p 251).
M.A.T.B.MN. : -.-.-.-.-.   First record: Lousley, 1940.

FLOWER INDUSTRY: (*top*) flower 'squares' enclosed by hedges and fences near Halangy, Mary's; (*bottom*) *Narcissus x medioluteus* growing in brambles on a field wall near Old Town, St Mary's

A native of South Africa, Mock Shamrock (as it is known in Scilly) is now thoroughly naturalised, and often a pest, in many parts of the world, including Bermuda, Malta, and countries bordering the Mediterranean. The bulb is annual and as it shrivels the food reserves pass into the tuberised root, and then into a new bulb and bulbils. It is these bulbils which facilitate the rapid spread; in Malta they are said to be wind-carried and this is probably the case also in Scilly. Dumping plants cleared from the fields has established it in stone-walls where it is almost impossible to eradicate.

I first noticed *O. pes-caprae* in 1938 in St Mary's, where it was near houses and possibly an escape from gardens. Early records were in the area bounded by Hugh Town, Old Town, Parting Carn and Carn Friars in which it is now in great abundance. It is all too plentiful over most of this island, Tresco, and St Martin's; more local on St Agnes, and rare on Bryher from which the first record was in 1967.

O. latifolia Kunth   Alien.

St Agnes: on a rubbish tip near the pool, 1963, D. McClintock. Tresco: field above Pool Road, 1963, D. McClintock; common weed in one part of Tresco Gardens, 1969 (spec.) R. Lancaster. On several occasions I have failed to find the plant from directions supplied by Mr McClintock.

ACERACEAE

Acer L.

A. pseudoplatanus L.   Sycamore.   Denizen.

Frequent.   Planted and also self-sown about buildings, by lanes, etc.

M.A.T.-.MN. : -.-.-.-.   First record: *c* 1650, ex Turner, 1964

Sycamore was introduced as a windbreak and seems to have spread from Holy Vale, St Mary's. It was planted there about 1650 (old manuscript quoted by Turner, *Scillonian* **159**, 154, 1964) and by 1822 there were 'fine trees' (Woodley, 1822, 220). In Tresco it was one of the trees planted to provide early shelter for the Abbey Gardens and it is still plentiful in Abbey Wood. In these two islands it reproduces freely from seed and is now frequent. On St Agnes I know it only from the Parsonage Garden, and on St Martin's from Higher Town.

AQUIFOLIACEAE

Ilex L.

I. aquifolium L.   Holly.   Alien.

M.-.T.-.-. : -.-.-.-.E.   First record: this *Flora*.

St Mary's: several old trees around one field, probably relics of a hedge, 1970, R. M. Burton. Tresco: one isolated bush on Appletree Banks, perhaps bird sown, 1970, R. M. Burton. Eastern Isles: Great Ganilly, one bush, 1970, P. MacKenzie. Although native and common in Cornwall, the present evidence suggests that in Scilly it is now starting to spread from bushes planted on St Mary's.

CELASTRACEAE

Euonymus L.

E. japonicus Thunb.   'Euonymus'.   Planted alien.

Common in fences as a windbreak, but also found apparently feral on cliffs and elsewhere. (Fig 9, p 185).   M.A.T.B.MN. : -.-.-.-.-.

First record: J. G. Owen in *Faire Lyonesse*, 1897.

A native of Japan, this shrub is commonly planted and its use as a windbreak probably dates back to about 1860. Plants in uncultivated places competing with native vegetation sometimes originate from uprooted hedges being thrown out of the bulbfields. *E. japonicus* flowers only in the mildest parts of Britain including Scilly.

E. europaeus L. 'Introduced as a fence plant' according to Paton (1968, p 54). Spindle, native on calcareous soils, would be useless as a windbreak and the record is an error.

LEGUMINOSAE

Lupinus L.

L. arboreus Sims.   Tree Lupin.   Naturalised alien.

Self-sown on dunes and sandy places.   Local.

M.-.T.B.MN. : -.-.-.-.-.   First record: Lousley, 1967.

In recent years this species has become established in sandy places and especially on Tresco, about the Block House, and on St Martin's, on the dunes at Laurence's Bay and Lower Town.

Ulex L.

U. europaeus L.   Gorse.   Native.

Rough cliff slopes, downs and hedgerows.   Abundant.

M.A.T.B.MN. : S.-.H.TN.E.   First record: Millett, 1852.

Gorse was formerly used for fuel, and charred wood has been found in Second and Third Century remains (Ashbee, 1955).

It was planted extensively by Augustus Smith 'about all the islands' for shelter and especially to protect young trees on the downs from wind and grazing animals. (Dorrien-Smith, *Scillonian*, **29**, 64, 1954).

U. gallii Planch.   Western Gorse.   Native.
Cliff slopes and downs.   Very common.
M.A.T.B.MN. : -.-.H.-.-.   First record: Ralfs, 1879.
This often grows mixed with *U. europaeus*, and is usually a shorter plant. It flowers from July to September and is in fruit when the common species blooms in March, April and May. A state of wind-swept situations, which is very well marked on St Helen's, has been recorded as var. *humilis* Planch. by C. E. Salmon in Davey, 1909.
U. minor Roth (*U. nanus* R. F. Forst.). Error. '*U. nanus* Forst.—very plentiful, Mr Smith; though I overlooked it' (Townsend, 1864). 'This is undoubtedly *U. gallii*, which is common' (Ralfs in Marquand, 1893). A specimen gathered by A. Somerville from St Mary's in 1890, confirmed by Arthur Bennett as *U. minor*, and now at Paisley Museum, is clearly small *U. gallii* from a windswept place. The same applies to a specimen in Herb. Dorrien-Smith which was named as *U. minor* by Edinburgh Vyvyan, 1953, 156). *U. minor* is a species of eastern distribution in England and unlikely to occur in Scilly.

Sarothamnus Wimm.
S. scoparius (L.) Wimm.   Broom.   Native.
Hedgebanks, quarries and slopes on granite.   Local.
M.-.T.B.-. : -.-.-.-.-.   First record: Millett, 1852.
St Mary's: roadside between Hugh Town and Old Town; Buzza Hill; The Garrison; Pelistry; Holy Vale; Salakee Downs. Tresco: locality not noted. Bryher: several places about the centre of the island; Pool; under Watch Hill.
This is all subsp. *scoparius*.

Ononis L.
O. repens L.   Rest Harrow.   Native.
Consolidated dune and a hedgetop.   Very rare.
M.-.-.-.MN. : -.-.-.-.-.   First record: Lousley, 1939.
St Mary's: in quantity on the top of a hedge near Borough, 1939 (308), two patches, 1970. St Martin's: The Plains, Herb. D.-S., also Grose, 1939 and still there 1970.

Medicago L.

M. sativa L.   Lucerne.   Established alien.

Cultivated fields, a roadside, and on a low wall.   Rare.

M.A.T.-.-. : -.-.-.-.-.   First record: Lawson, 1870.

St Mary's: Beeby, 1873; Ralfs, 1879. St Agnes: near Middle Town. Tresco: Lawson, 1870; cultivated fields, Herb. D.-S.; roadside near cottages, Old Grimsby, 1939 onwards.

The Old Grimsby locality seems to be permanent and may be the one known to Lawson nearly a century ago. Lucerne is still in regular cultivation.

M. lupulina L.   Black Medick.   Native.

Cultivated fields and gardens.   Rare.

M.A.T.-.MN. : -.-.-.-.-.   First record: Townsend, 1864.

Townsend gave this as 'common', and it may have been so as about that time it was sown in clover mixtures on barley land (Scott & Rivington, 1870). Recent records have been few, mostly near houses and usually in small quantity, but it is plentiful in Tresco Gardens.

M. polymorpha L. (*M. hispida* Gaertn.).   Toothed Medick. Native.

Sandy banks by the sea, walls and bulbfields.   Local.

M.A.T.B.MN. : -.-.-.-.-.   First record: Ralfs, 1879.

St Mary's: Smith, 1909; Old Town Bay, growing in the sand, Perrycoste in Davey, 1909; bulbfield between Hugh Town and Old Town (109); wall near Parting Carn (473); Porth Cressa, Grose. St Agnes: Priglis Bay; garden of Turk's Head Inn; Middle Town, Grose. Tresco: Old Grimsby, Tellam in Ralfs, 1879 and Smith, 1909. Bryher: Town Bay. St Martin's: Middle Town; common below School.

The common plant in Scilly is var. *vulgaris* (Benth.) Shin. (*M. denticulata* Willd.) with spiny pods, and this I think is likely to be native on sandy ground near the sea from which it has spread to bulbfields. Plants without spines on the fruits—var. *brevispina* (Benth. emend.) Heyn (*M. apiculata* var. *confinis* Koch)—occurred in a bulbfield between Hugh Town and Old Town, St Mary's (108). Intermediates with very short spines (*M. apiculata* Willd.) have been found:—St Agnes: flower-pot in Turk's Head Inn (437); path by Priglis Bay, Wanstall. St Martin's: bulbfield, Middle Town (110).

M. arabica (L.) Huds.   Spotted Medick.   Native.
Pathsides and a pest in gardens and bulbfields.   Very common.
M.A.T.B.MN. : -.-.-.-.-.   First record: Townsend, 1864.

Melilotus Mill.
M. officinalis (L.) Pall. ˙ Common Melilot.   Casual.
Sandy ground near the shore.   Very rare.
-.A.T.-.-. : -.-.-.-.-.   First record: Millett, 1852.
St Agnes: sandy flat near The Gugh, 1933, Herb. D.-S. Tresco:
New Grimsby near the beach, Townsend, 1864—of this record
Ralfs, 1879 wrote 'I believe this is *M. arvensis*'.
The only specimen I have seen is the one at Tresco Abbey which
has no fruit, but which I think is correctly assigned to this species.

Trifolium L.
T. ornithopodioides L.   Birdsfoot Fenugreek.   Native.
Tracksides, short turf on greens and cliffs.   Common.
M.A.T.B.MN. : -.-.-.-.-.   First record: Townsend, 1864.
The main flowering period is in May and June, but in wet seasons
this is much extended and I have found flowers as late as September.

T. pratense L.   Red Clover.   Native?
Waysides and pastures—sometimes planted as fodder.   Common.
M.A.T.B.MN. : -.-.-.-.-.TN.   First record: Townsend, 1864.
Like Lucerne, Red Clover is an old introduction as a fodder plant.
I have seen it only in man-made habitats on the five inhabited
islands, with the addition of Tean which was formerly cultivated.
It may be native but I have no clear evidence of this.
T. medium L. Error. This was recorded by Boyden in Tonkin & Row,
1893. Several observers have recorded it from St Agnes but I have been
sent specimens of *T. pratense* from that island which closely resemble
T. medium superficially and it is likely that Boyden was similarly deceived.
It is rare on islands round the British coast and not very likely to occur.

T. incarnatum L.   Crimson Clover.   Planted and casual alien.
Clover-leys and occasionally roadsides.   Rare.
M.-.T.-.-. : -.-.-.-.-.   First record: Scott & Rivington, 1870.
St Mary's: occasionally planted—e.g. in fields by Old Town Lane.
Tresco: field near New Inn; field near Dial Rocks.

T. arvense L.   Haresfoot Clover.   Native.
Sandy places, and on walls.   Frequent.
M.A.T.B.MN. : S.-.-.-.-.   First record: Millett, 1852.

T. striatum L.   Soft Trefoil.   Native.

Sandy places.   Rare.

M.A.T.B.MN. : -.-.-.-.   First record: Townsend, 1864.

St Mary's: 1872, Beeby in Herb. S.L.B.I.; Ralfs, 1879. St Agnes:
Ralfs, 1879; Dallas (117). Tresco: Ralfs, 1879; Abbey grounds—
very fine plants (118). Bryher: near Samson Hill. St Martin's:
Ralfs, 1879.

Downes in his copy of Davey, 1909 recorded 'var.' *erectum* Leighton
from Tresco in a manuscript note. This is a tall branched state,
and he probably had in mind specimens similar to mine from
Abbey grounds.

T. scabrum L.   Rough Trefoil.   Native.

Shingle and sandy upper beaches, also by tracks and on shallow
soil over granite.   Rare.

M.A.T.B.MN. : -.-.-.-.   First record: Ralfs, 1879.

St Mary's: Beeby in litt to Townsend, 1873 (this is the earliest
reference but he only collected *T. striatum*, see above); Ralfs, 1879;
Porthellick Bar, very small (300); ramparts of Star Castle (418).
St Agnes: Ralfs, 1879; near the quay, Grose. Tresco: Ralfs, 1879;
Herb. D.-S. Bryher: 'on beach where you land', Dallas (121);
Northward. St Martin's: Lower Town, Herb. Dallas; sandy track,
Higher Town Bay (696).

T. subterraneum L.   Subterranean Clover.   Native.

Cliff slopes, greens, roadsides, walls and sandy flats.   Very
common.

M.A.T.B.MN. : -.-.-.-.   First record: Townsend, 1864.

T. strictum L. 'I feel pretty sure of having noticed this but omitted to
check it', Townsend *MS.*, 1862. Growing in Cornwall and the Channel
Isles this is a likely species to occur, but the record is too uncertain for
acceptance.

T. glomeratum L.   Clustered Trefoil.   Native.

Sand-dunes, tracks, sandy cultivated ground, a quarry, and wall-
tops.   Frequent.

M.A.T.-.MN. : -.-.-.-.   First record: Townsend, 1864.

St Mary's: earthwalls, 1862, Townsend in Herb. Townsend; dry
banks, 1872, Beeby in Herb. S.L.B.I.; hedgebanks, 1876, Curnow
in Herb. Tellam; hedgebanks, not unfrequent, Ralfs, 1879; 1898,
A. G. Gregor in Herb. S.L.B.I. St Agnes: 1877, Tellam in Herb.

Tellam; facing towards The Gugh, very fine, Ralfs, 1879. Tresco: Old Grimsby, Ralfs, 1879; very common in Abbey Grounds (115); small quarry in Racket Town Lane (456); track near Plumb Hill. St Martin's: Ralfs, 1879; shore by Lower Town (114).

T. suffocatum L.   Suffocated Trefoil.   Native.
Tracksides and sandy ground, and especially on greens.
Common.
M.A.T.B.MN. : -.-.-.-.-.   First record: Ralfs, 1879.
This species is too common to deserve detailed localities, but is easily overlooked on account of its very early flowering. In normal seasons it dries up and disintegrates in May.

T. hybridum L.   Alsike Clover.   Casual.
Sown in clover-leys and sometimes appearing in other crops the following year; also roadsides, not persisting.   Rare.
M.A.-.-.MN. : -.-.-.-.-.   First record: Somerville, 1893—without locality.
St Mary's: 1938, M. Knox; field near Parting Carn; roadside west of Rosehill, abundant. St Agnes: oatfield, Middle Town, Grose. St Martin's: 1938, M. Knox.

T. repens L.   White Clover, Dutch Clover.   Native.
Roadsides, coastal slopes, walls, dunes, etc.   Common.
M.A.T.B.MN. : S.-.H.TN.-.   First record: Townsend, 1864.
var. *townsendii* Beeby. This handsome clover with rich purple flowers impresses many botanists visiting Scilly. The colour is very different from that of the reddish-purple forms of *T. repens* found on the mainland, and it is often uniform over large colonies although intermediates may also be found. It is tempting to suspect that the strong development of colour pigment, which is often seen in the leaves as well as the flowers, may be associated with the high actinic content of the sunlight which causes so many sunburn casualties among visitors. On the other hand, plants with white flowers sometimes grow close to those with purple ones, and the latter are said to be constant in cultivation when grown in other parts of England and Scotland (e.g. J. T. Boswell Syme at Balmuto in August 1883 in Herb. Mus. Brit., and see N. E. Brown, 1892, *English Botany—Supplement* p 63). Experimental work on this variant is very desirable.
Townsend was the first to bring this to the notice of botanists

(*J. Bot.*, **1**, 216, 1863) having found it on wet sandy flats near Tresco pool. The following year it was fully described and illustrated in colour by Babington, but he did not give it a name (*J. Bot.*, **2**, 1–3, *t.* 13, 1864). It seems that the correct varietal citation is *T. repens* var. *townsendii* Beeby in *Rep. B.E.C.* **1872–4**, 14, 1875 which is validated by reference to Babington's account—I am grateful to B. L. Burtt for assistance on the question of nomenclature. No specimens collected by Townsend can be found in his herbarium.

Var. *townsendii* is most plentiful on Tresco, Bryher, St Martin's and Samson, but it also occurs on St Mary's, St Helen's and Tean.

T. occidentale D. E. Coombe.   Native.
Shallow soils on slopes over granite; also on blown sand.
Common. (Fig 7, p 125.)
M.A.T.B.MN. : -.-.-.-.-.   First record: Vercoe in Coombe, 1961.
St Mary's: Peninnis Head, short turf on granite, Vercoe in Coombe, 1961; Peninnis Head, 1963 and 1967 (890); Salakee Downs. St Agnes: short turf on granite, also on Gugh, Vercoe in Coombe, 1961. Tresco: abundant near the Blockhouse, both on dune sand and on granite, 1966, Coombe. Bryher: towards Shipman Head, 1963; Northward, 1963 (854). St Martin's: Brandy Point, 1967 (873).
This new species, related to *T. repens*, was described by D. E. Coombe in 1961 (*Watsonia*, **5**, 68–87, 1961) and he published a supplementary account six years later (*Watsonia*, **6**, 271–275, 1967). It is known with certainty only from west Cornwall, the Isles of Scilly, the Channel Isles, the coast of Brittany and the Cotentin peninsula in north-west France. It is more resistant to high concentrations of salt from sea-spray than *T. repens*, and tends to flower earlier. No doubt it is more common in Scilly than these few records indicate: no search for it has yet been made on the uninhabited islands.

T. fragiferum L.   Strawberry Trefoil.   Native.
Sand-dunes and sandy garden soil   Rare.
-.-.T.-.MN. : -.-.-.-.-.   First record: Lawson, 1870.
Tresco: Lawson, 1870; Abbey Grounds; Old Grimsby. St Martin's: 1904, Reid in Davey, 1909; Lower Town.

T. campestre Schreb. (*T. procumbens* auct.).   Hop Trefoil.
Native.
Sand-dunes, roadsides and bulbfields.   Frequent.
M.A.T.B.MN. : -.-.-.TN.-.   First record: Townsend, 1864.
Exceptionally small plants occur in consolidated turf on the dunes.

T. dubium Sibth.   Common Yellow Trefoil.   Native.
Cultivated ground, roadsides, etc.   Very common.
M.A.T.B.MN. : -.-.-.TN.-.   First record: Townsend, 1864.

T. micranthum Viv. (*T. filiforme* nom. ambig.)   Lesser Yellow
Trefoil.   Native.
Sandy tracksides and greens.   Rather rare.
M.A.T.B.MN. : -.-.-.TN.-.   First record: Townsend, 1864.

Anthyllis L.
A. vulneraria L.   Kidney Vetch.   Native.
Extinct.   Habitat not recorded.   First record: Millett, 1852.
This was included in the list published by the Misses Millett, who
were very reliable, and it is mentioned by Ralfs in a letter to
Townsend in 1876. Ralfs gave it in his *Flora*, 1879 as 'common'
but many of his statements of frequency were wildly inaccurate.
No recent worker has seen it and it seems likely that the species
was always very rare and is now extinct. Kidney Vetch is plentiful
on the cliffs of Land's End, and occurs in some of the Channel
Isles, and its present absence from Scilly is surprising.

Lotus L.
L. corniculatus L.   Common Birdsfoot Trefoil.   Native.
Walls, roadsides, heaths, cliffs, etc.   Very common.
M.A.T.B.MN. : S.AT.H.TN.E.   First record: Millett, 1852.
Townsend, 1864 recorded var. *crassifolius* Pers. as one of the most
common forms. This is a state with thick fleshy leaves developed
in response to heavy concentrations of salt, and it is common on
the cliffs and particularly well marked on the small spray-swept
islands. He also gives 'var. *villosus* Ser.' from St Martin's. In places,
such as the slopes of Bryher, *L. corniculatus* is so abundant that
from a distance it colours the hills yellow.

L. uliginosus Schkuhr   Marsh Birdsfoot Trefoil.   Native.
Lanesides, rough cliff slopes and marshes; characteristic of wet

places but by no means confined to them.   Very common.
M.A.T.B.MN. : S.-.H.TN.E.   First record: Townsend, 1864.

L. hispidus Desf. ex DC.   Lesser Birdsfoot Trefoil.   Native.
Walls, bare places on cliffs and carns, and cultivated fields.   Very
common.
M.A.T.B.MN. : -.-.-.TN.-.   First record: Townsend, 1864.
L. hispidus occurs in two types of habitat with different forms.
In natural habitats, such as shallow soils on granite carns, where
it is often associated with *Ornithopus pinnatus*, and dunes, the
plants are usually small, with stems only 5–6 cm, and shaggy with
long hairs. In bulbfields and other cultivated ground to which the
species has spread from natural habitats, the plants are usually
with ascending branches up to 50 cm or more long, with larger and
less shaggy leaves.
L. angustissimus L. has been much confused with the last species and
all the records are likely to be errors. Curnow stated that it grew 'in
many places' (*Hardwick's Science Gossip*, **1876,** 162) but the only sheet
of this species in his herbarium is from Mousehole near Penzance,
and he added a note 'I never gathered it since'. Boyden included it in
his 1890 and 1893 lists but if by '*L. pilosus*' he intends the very common
L. uliginosus, then he fails to account for *L. hispidus* which is also common. Ralfs, writing to Townsend in 1876, said that *L. angustissimus*
grew in St Mary's and Tresco, but was less common than *L. hispidus*.
Vyvyan 1953, cites a specimen in Herb. Dorrien-Smith as determined
by Edinburgh as *L. angustissimus*, but this is *L. uliginosus*. There are
correctly named specimens in Herb. Townsend labelled Tresco, 1845,
but Townsend did not visit Scilly until 1862 and the sheet also carries
material of *L. hispidus*. Many recent visitors to Scilly have reported
L. angustissimus but their specimens have always turned out to be
L. hispidus, and until proper evidence can be produced the species cannot
be accepted.

Ornithopus L.
O. perpusillus L.   Common Birdsfoot.   Native.
On thin soil over granite on heaths and carns where there is little
competition, and on sand in cultivated fields and dunes.   Common.
M.A.T.B.MN. : S.AT.-.TN.E.   First record: Millett, 1852.
Townsend, 1864, says 'it occurs frequently with the pod and
calyx quite glabrous' but plants in his herbarium from St Agnes
labelled 'plants wholly glabrous' have a few hairs on the pods and
calices. The pubescence varies widely and I have had it with hairy
pods (809), and with pods glabrous and the calyx hairy (731).
Wholly glabrous plants (var. *glaber* Corb.) are rare.

O. pinnatus (Mill.) Druce (*O. ebracteatus* Brot.)   Jointed Birds-
foot.   Native.
On shallow soil on granite carns, heathy slopes, sandy tracks, and
amongst short grazed Ling. Often associated with *Lotus hispidus*.
Frequent. (Fig 7, p 125; Map 8, p 172.)
M.A.T.B.MN. : -.-.-.TN.E.

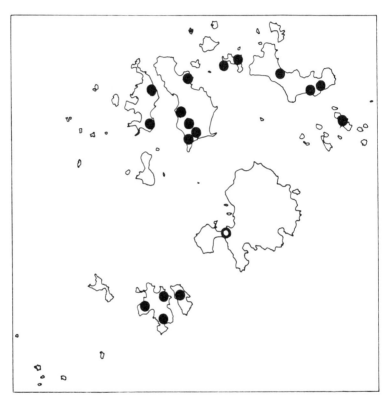

Ornithopus pinnatus, Jointed Birdsfoot

First record: 'Rev. H. Penneck in a letter to Dr Barham, published
in *J. Roy. Inst. Cornw.*, 1838' (Davey, 1909).
St Mary's: Near Hugh Town, Tellam ex Ralfs, 1877A and Herb.
Tellam. St Agnes: Not uncommon, Ralfs, 1879; The Gugh,
Downes in his copy of Davey, 1909; near Troy Town, Dallas;
Carn Grigland (731). Tresco: May, 1838 and 1839, Miss White,

Herb. Kew; Plumb Hill and Carn Near, Herb. D.-S.; Appletree Banks (101); Abbey Grounds (102); Merchant's Point (703). Bryher: 1876, Curnow, Herb. Tellam; Northward (103); Samson Hill (706). St Martin's; Above Great Bay (104); near Brandy Point Carn (734). Tean: East side, Herb. Dallas; hill facing St Martin's (707); Old Man (705). Eastern Isles: Great Ganilly, M. Jaques. Many records which add nothing to the history or distribution have been omitted.

O. pinnatus was discovered by Miss Matilda White in Tresco at the end of April, 1838 as described in detail in her letter to Sir W. Jackson Hooker dated December 17, 1938 (*Hooker Correspondence*, Kew). Her specimens varied from a few inches in length to half a yard, and the plant grew with heath species on a granite hill away from the Abbey. *O. pinnatus* occurs also in Guernsey, Alderney, Sark and Herm (but not Jersey) and is unknown elsewhere in the British Isles. Its flowering period extends from April to October but its habitats dry out very quickly, and its appearance depends on moist conditions continuing sufficiently long to enable it to mature. I have failed to refind it on St Mary's.

Vicia L.
V. hirsuta (L.) Gray   Hairy Tare.   Native.
Hedges, waste ground, and field borders.   Frequent.
M.A.T.B.MN. : -.-.-.-.-.   First record: Millett, 1852.

V. tetrasperma (L.) Schreb.   Smooth Tare.   Native.
Hedges, waste ground, and arable fields.   Locally rather frequent.
M.A.-.-.MN. : -.-.-.-.-.   First record: Townsend, 1864.
I have failed to find this species on Tresco and Bryher, and in the other inhabited islands it is less common than *V. hirsuta*.

V. cracca L.   Tufted Vetch.   Native.
Roadsides, field borders and grassy places.   Frequent.
M.A.T.B.MN. : -.-.-.TN.-.   First record: Millett, 1852.
Most of the habitats are disturbed ground but Mr Grose found it in 1952 on Old Man, an islet off Tean.
V. sepium L. Recorded by Somerville, 1893, without locality and it has been listed for St Mary's and St Agnes, but confirmation is required.

V. sativa L.   Common Vetch.   Naturalised alien.

Waysides, field borders, weed in bulbfields, marshes and rough ground.   Common.

M.A.T.B.MN. : -.-.-.TN.-.   First record: Townsend, 1864.

V. sativa grows mainly in places where it may have originated from crops but it also occurs where there is no evident association with present cultivation. For example, it grows in bracken round the ruins on Tean (710) on ground unlikely to have been cultivated for a century.

V. angustifolia L.   Narrow-leaved Vetch.   Native.

Dry sandy or heathy places, in sandy bulbfields and in Ammophiletum on dunes.   Common.

M.A.T.B.MN. : -.-.H.TN.E.   First record: Townsend, 1864.

A very variable species in Scilly. A slender form about 60 cm tall was provisionally given by Townsend, 1864 as *V. uncinata* Bor., and similar plants still grow on St Martin's amongst *Ammophila* in Great Bay (692), and on St Helen's in bracken near the Pest House (692). On The Gugh there is a large-flowered form which, as Dr R. C. L. Burges has pointed out, resembles the Jersey plant which G. C. Druce described as var. *garlandii*. It seems unlikely that these merit taxonomic distinction, but the group needs to be studied on modern lines.

V. lathyroides L. Tresco: Appletree Banks, May 1971, J. E. Raven. This occurs in the Channel Isles but the record is new to south-west England.

V. bithynica (L.) L.   Bithynian Vetch.   Native?

Roadside and cultivated ground.   Rare and very local.

M.-.-.-.-. : -.-.-.-.-.   First record: Downes, 1924 (see below).

Cultivated ground, Scilly Isles, June 1923, Harold Downes, *Rep. B.E.C.*, **7**, 381, 1924 and Herb. Mus. Brit. St Mary's: grassy roadside between Hugh Town and Rocky Hill, 1958, R. P. Bowman (spec.); bulbfield near Trewince, 1967, P. Z. MacKenzie (spec.).

This species is accepted as native on The Lizard and may be so in Scilly in spite of the nature of the habitats. It has been established that no peas or beans had been planted in the field at Trewince since before 1939.

Lathyrus L.

L. pratensis L.   Meadow Vetchling.   Native.

Roadside verges, etc.   Very rare.

M.A.T.-.-. : -.-.-.-.   First record: Lousley, 1939.

St Mary's: Roadside near Newford; High Cross Lane, Dallas. St Agnes: 1967, H. M. Quick. Tresco: The Duckery, under south Terrace of the Abbey, A. A. Dorrien-Smith.

L. sylvestris L. Townsend, 1864, says he saw one plant which was hardly in flower. It has not been reported since, and there are no habitats in Scilly where I would expect it to grow. It is possible that Townsend confused it with the cultivated Everlasting Pea, *Lathyrus latifolius*, which is grown in the islands and may be seen, for example, on garden walls in Hugh Town.

ROSACEAE

Filipendula Mill.

F. ulmaria (L.) Maxim. (*Spiraea ulmaria* L.)   Meadowsweet. Native.

In the marshes.   Extinct.   M.-.-.-.-. : -.-.-.-.

First record: Lawson, 1870—without locality.

St Mary's: Marshes, Ralfs, 1879.

This species has not been reported for 90 years and is probably extinct. Its extreme rarity in Scilly, where there are plenty of suitable habitats, is a great contrast to its abundance in Cornwall. In the Channel Isles it occurs only in Jersey, and there it is very rare.

Rubus L.

R. idaeus L.   Raspberry.   Garden outcast.

St Mary's: Garrison Hill—'in scrub beyond tennis courts', April 1964, Mrs Mary Briggs.

R. caesius L.   Dewberry.   Native.

Rough cliff slopes.   Local.

M.-.-.-.MN. : -.-.-.-.   First record: Vyvyan, 1953.

St Mary's: fresh specimen named by F. Rilstone, 1945, collector not known; shore at Bar Point, 1953; cliff path from Halangy to Bar Point, 1963; near Innisidgen Carn, 1963. St Martin's: Cruther's Point, 1967. The specimen cited by Vyvyan from Herb. Dorrien-Smith, which was named at Edinburgh, probably came from Tresco. The species is no doubt more frequent than these records indicate.

R. fruticosus L. *sensu lato.*   Blackberry, Bramble.   Native.
Roadsides, heaths, rough cliffs, consolidated dune, etc.
Abundant.
M.A.T.B.MN. : S.AT.H.TN.E.   First record: Millett, 1852.
No leading batologist has visited Scilly and very little reliable
information is available about the microspecies which occur other
than the abundant and distinct *R. ulmifolius*. Townsend, in 1864,
recorded five species as determined by Prof. C. C. Babington but
his specimens cannot be found, and the names cannot be related
to those now in use. I and others have done our best to obtain up
to date names by collecting material but unfortunately W. C. R.
Watson and F. Rilstone passed away and E. S. Edees and J. E.
Woodhead felt that they were not sufficiently acquainted with the
brambles of Cornwall and the Channel Isles to name more than a
few of the taxa represented. More recently, B. A. Miles has named
a few sheets for myself and Miss K. Marks. Thanks to the help of
these batologists it is possible to give the following records but no
adequate account will be forthcoming until an experienced worker
has studied these critical plants in the field.

R. myriacanthus Focke—E. S. Edees gave this name to a gathering
from School Green, Tresco (702). Another sheet belonging to the
section *Triviales* came from Bar Point, St Mary's (836).

R. riddelsdellii Rilstone—St Mary's: Halangy Point (675) det.
B. A. Miles; near Rocky Hill, K. Marks, det. B. A. Miles.

R. ulmifolius Schott (*R. rusticanus* Merc.)   Native.
Beneath bracken in the characteristic community of the cliff
slopes, dunes, and marshes.   Abundant.
M.A.T.B.MN. : S.-.H.-.E.   First record: Somerville, 1893.
This is the common abundant bramble of Scilly. Out of a total of
35 gatherings of Rubi selected by three collectors for maximum
variety no fewer than 27 were determined by experts as *R. ulmifolius*.
Unlike the other species it is not apomictic, and hence it is very
variable. W. C. R. Watson determined five gatherings made by
O. Polunin as follows:
     ssp. *vulgatus* Sud. St Mary's: Peninnis Head, among *Ulex* (641).
     ssp. *anisodon* Sud. St Agnes (640). var. *cannabinus* (Boul. and
         Let.) St Mary's: Old Town (639).

ssp. *insignitis* (Timb. and Muell.) Sud. St Mary's: Marsh by wall in swampy ground, Old Town (638).

ssp. *dilatifolius* Sud. St Mary's: Peninnis Head in an exposed place on the Outer Head (637).

R. pseudadenanthus W. C. R. Watson (*R. adenanthus* Rogers non Boul. and Gill.) Tresco: Borough, in pine wood, Polunin (165) det. W. C. R. Watson.

R. hastiformis W. C. R. Wats. St Mary's: Low Pool, Rose Hill (660); in bracken, Carn Morval Down (672)—both det. B. A. Miles.

R. cornubiensis (Rog. and Ridd.)Rilst.   St Mary's: Holy Vale, K. Marks, det. B. A. Miles.

Potentilla L.
P. sterilis (L.) Garcke   Recorded by Tellam in Davey, 1909 without locality. In the absence of confirmation this must be treated as an error.

P. anserina L.   Silverweed.   Native.
Waysides, gateways, waste places and beaches.   Common.
M.A.T.B.MN. : -.-.H.TN.-.   First record: Millett, 1852.
Occasionally, as on Tean, with the leaves silvery on the upper surface as well as below.

P. erecta (L.) Rausch.   Tormentil.   Native.
Heathy places on the downs and cliffs.   Abundant.
M.A.T.B.MN. : S.-.H.TN.E.   First record: Millett, 1852.
x *reptans* (*P.* x *italica* Lehm.). St Mary's: roadside wall near Maypole (745) det. D. E. Allen.

P. anglica Laichard. (*P. procumbens* Sibth.). Trailing Tormentil.
Native.
On sides and tops of the earth-filled stone hedges.   Rare.
M.-.T.B.-. : -.-.-.-.-.   First record: Somerville, 1893—without locality.
St Mary's: Near Tremelethen (127); near Old Town; Maypole.
Tresco: Herb. D.-S. Bryher: on stone hedge near Vine Inn.
These records may require revision. The concept of this species has changed in recent years and my only specimen is not typical.

P. reptans L.   Creeping Cinquefoil.   Native.
Walls, waysides and waste places.   Frequent.
M.A.T.B.MN. : -.-.-.-.-.   First record: Townsend, 1864.

Sibbaldia L.
S. procumbens L. Recorded by Vyvyan, 1953 from a specimen in Herb. D.-S. determined by the Royal Botanic Gardens, Edinburgh. The species is only known in Britain from mountains in the north and the record cannot be taken seriously.

Fragaria L.
F. vesca L. Wild Strawberry. This was recorded by Tellam in Davey, 1909 without locality. I noted it in 1956 as a weed in Abbey Gardens, Tresco but here it was an obvious introduction and in view of the numerous errors of transcription in Davey, it cannot be accepted for Scilly without confirmation.

Geum L
G. urbanum L.   Herb Bennet.   Native.
Recorded by Boyden in Tonkin and Row, 1893, and by Tellam in Davey, 1909—in both cases without locality. As these are independent records of a species unlikely to be confused they must be accepted, but efforts to refind this shade-loving plant have been unsuccessful.

Aphanes L.
A. arvensis L.   Parsley Piert.   Native.
Bulbfields.   Rare.
M.-.T.-.MN. : -.-.-.-.-.   First record (*sensu stricto*): Lousley, 1967.
St Mary's: bulbfield weed, Old Town (808), 1957; bulbfield, Old Town Lane (657), 1954; field behind Porthloo, B. M. C. Morgan. Tresco and St Martin's: B. M. C. Morgan.

A. microcarpa (Boiss. and Reut.) Rothm. Parsley Piert. Native.
Cultivated fields, sandy grounds, and cliffs.   Very common.
M.A.T.B.MN. : S.-.-.-.E.   First record: Townsend, 1864—as *Alchemilla arvensis*.
This is the common Parsley Piert of Scilly and many specimens have been determined for me and for others by Dr S. M. Walters.

Rosa L.
R. multiflora Thunb. This is in hedges by the drive to Salakee Farm, St Mary's, where it was planted, and also in Porthellick Marshes.

R. canina L.   Dog Rose.   Native.

Lanesides, rough places on cliffs, consolidated dune.   Rare.

M.-.T.-.MN. : -.-.-.-.   First record: Townsend, 1864.

St Mary's: Single specimens in Heugh Town Marshes, Townsend, 1864; 1 bush near Maypole. Tresco: Middledown, Herb. D.-S.; 2 bushes on Appletree Banks; 2 at Gimble Porth, several on Monument Hill. St Martin's: 3 bushes, south coast, Grose; 1 bush on cliffs near Brandy Point; 1 near Turfy Hill; 3 bushes in *Ammophila*, Higher Town Bay.

It is remarkable that less than twenty bushes of Dog Rose have been found in Scilly as the species is very common in Cornwall.

R. rubiginosa L.   Sweet Briar.   Perhaps planted.

Consolidated dune.   Very rare.

-.-.T.-.-. : -.-.-.-.   First record: Vyvyan, 1953.

Tresco: Penzance Road Gate (Appletree Banks), Herb. D.-S.; still there, 1957, 1967, and another bush opposite the Garden Gate.

Prunus L.

P. spinosa L.   Blackthorn.   Native.

Hedges and cliff fields.   Rare.

M.-.T.-.-. : -.-.-.-.   First record: Millett, 1852.

St Mary's: Ralfs' *Flora;* hedge near Newford; above Halangy Point, 1952, Grose. Tresco: 'cliff fields, walls and Benzies', Herb. D.-S.; Plumb Hill.

P. domestica L. subsp. *institia* (L.) Poir. Bullace. Recorded by Somerville, 1893 without locality and confirmation required.

P. cerasifera Ehrh.   Cherry Plum.   Alien, planted.

St Mary's: in a hedge near the Airport, 1953, D. McClintock.

Crataegus L.

C. monogyna Jacq.   Hawthorn.   Native.

Hedges and bushy places.   Local.

M.-.T.-.-. : -.-.-.-.   First record: Millett, 1852.

St Mary's: north of Porthellick Pool; Frequent about Holy Vale, Watermill Cove and Maypole; hedge near Salakee; one plant on Garrison; several plants near Bar Point; roadside near the Telegraph; Tregear's Porth. Tresco: one bush near Borough; one near Old Grimsby; two at Northward; one below Abbey Hill.

Malus Mill.

M. sylvestris Mill.   Crab Apple.   Probably alien.

Rough places and hedges.   Very rare.

M.-.T.B.-. : -.-.-.-.-.   First record: Lousley, 1967.

St Mary's: one bush on coast path north of Watermill Cove. Tresco: one tree, Monument Hill.

subsp. *mitis* (Wallr.) Mansf. (=*M. domestica* Borkh.). Bryher: in field hedge under Samson Hill, 1967, F. Russell Gomm.

These apples may have been planted, but are more likely to have originated from seed. All three are in places frequented by picnickers.

CRASSULACEAE

Sedum L.

S. anglicum Huds.   English Stonecrop.   Native.

Coastal rocks, heathy downs, sandy places, etc.   Abundant.

M.A.T.B.MN. : S.AT.H.TN.E.   First record: Millett, 1852.

The flowers in June are a colourful feature of the cliffs and the smaller islands.

S. album L.   White Stonecrop.   Alien.

Reported by R. P. Bowman from stone walls about Porthloo, St Mary's in 1957.

S. acre L.   Biting Stonecrop.   Native.

Sandy places near the shore.   Rare.

M.A.T.B.MN. : S.-.H.TN.-.   First record: Townsend, 1864.

St Mary's: Shore near Old Town, Towns.; shore at Porthellick. St Agnes: The Gugh, Ribbons and Wanstall. Tresco: Bathing-house and Farm, Herb. D.-S. Bryher: near shore north-east of Samson Hill. St Martin's: Lower Town. Samson, North Hill, Grose; east shore dune, A. Conolly. Tean: Grose. Northwethel.

Sempervivum L.

S. tectorum L. Houseleek. Recorded from Hugh Town, St Mary's, April 1890 by C. J. Plumtre in his notebook. No doubt planted as protection against lightning or for veterinary use.

Crassula L.

C. decumbens Thunb.   Established alien.

Weed in damp sandy bulbfields.   Locally plentiful. (Fig 8, p 155; Fig 12, p 282.)

M.-.-.-.-. : -.-.-.-.-.   First record: Lousley, 1960 (see below).
St Mary's: plentiful in bulb squares about Bant's Carn Farm and
Halangy, extending for about 300 yards.
This South African species was first noticed by Mr G. Baines in
bulbfields at Seaways in February 1959 and material supplied by
him was sent through Mr H. G. Morgan to Kew for identification.
Further material in fruit sent in April was identified by Mr R. D.
Meikle. It is common along the sides of tracks and probably spread
on the wheel of farm implements but has not greatly extended its
range. See Lousley in *Proc. B.S.B.I.*, **4**, 42–43, 1960.

Umbilicus DC.
  U. rupestris (Salisb.) Dandy   Wall Pennywort.   Native.
  Walls, rocks and hedges.   Abundant. (Plate p 193).
  M.A.T.B.MN. : S.AT.H.TN.E.   First record: Millett, 1852.
  This species is able to thrive, and grow exceptionally large leaves,
  in habitats with very high salt concentration. Thus it grows on
  Nor-nour, Great Innisvouls, Annet and other islands in crevices in
  rocks drenched in salt spray.

Aeonium Webb. and Berth.
  A. canariense (L.) Webb. and Berth. is reported by R. C. L. Howitt
  from St Agnes 'down by the bar'. What I believe to be *A. cuneatum*
  persists on the end of a building between Lower and Middle Town,
  St Martin's, where there are many plants which must be self-sown.
  I have not seen it in flower.

ESCALLONIACEAE

Escallonia Mutis ex L.f.
  E. macrantha Hook. and Arn.   Alien, planted as windbreak.
  Common. (Fig 9, p 185.)
  M.A.T.B.MN. : -.-.-.-.-.   First record: Lousley, 1939.
  E. macrantha is a native of the Isle of Chiloé, Chile, and was
  brought to England by William Lobb from his 1845 expedition
  and grown and distributed by Veitch of Exeter. This is likely to
  be the source of the plants grown in Tresco Abbey Garden. It
  was in this garden in 1873 (Hunkin, 1947), and no doubt earlier,
  and was in use as a windbreak in other islands before 1890 (Brewer,
  1890). The close, tough, evergreen leaves give effective protection
  and it has been extensively planted, and is often used in combination

with the taller growing Pittosporum. Since about 1950 it has been less commonly planted owing to the heavy demands it makes on the soil. I have no evidence that it spreads from seed, but bushes are found on cliffs or waste ground either as relics of old plantings or having rooted after being thrown out of bulbfields.

LYTHRACEAE

Lythrum L.

L. salicaria L.   Purple Loosestrife.   Native.
In marshes and by a pond.   Local.
M.-.T.-.-. : -.-.-.-.   First record: Millett, 1852.
St Mary's: Higher Moors, in quantity from Porthellick to Treme-lethen. Tresco: In Duckery, Herb. D.-S.; still there.

L. portula (L.) D. A. Webb (*Peplis portula* L.)   Water Purslane.
Pools and splashes which dry out in summer.   Rare.
M.-.T.-.-. : -.-.-.-.   First record: Millett, 1852.
St Mary's: Ralfs in litt. to Townsend, 1876; pond by roadside north of Normandy, 1956. Tresco: abundant around Abbey Pool; cart-track on Middle Down; Racket Town Lane (510).

THYMELAEACEAE

Daphne L.

D. laureola L.   Spurge Laurel.   Native.
In hedge.   Very rare.
M.-.-.-.-. : -.-.-.-.   First record: Lousley, 1967.
St Mary's: In thick hedge on right hand side of the road from Hugh Town to the Airport, near willows, 1954, R. C. L. Howitt. The occurrence of a single plant is in keeping with the pattern of distribution of this species in Cornwall where it is accepted as native.

ONAGRACEAE

Epilobium L.

E. hirsutum L.   Great Hairy Willowherb.   Native.
Ditches in marsh.   Very rare.
M.-.-.-.-. : -.-.-.-.   First record: Smith, 1909.
St Mary's: Old Town Marshes, rare, Smith; in a ditch behind the 'gardens' at Porth Mellon, 1967, P. Z. MacKenzie (864). This is at the north-west corner of Lower Moors (=Old Town Marshes) and may be the same as Smith's locality.

E. parviflorum Schreb.   Hairy Willowherb.   Native.
Margin of a pool, and weed in cultivated ground.   Very rare.
-.-.T.-.-. : -.-.-.-.-.   First record: Vyvyan, 1953.
Tresco: Tresco Gardens, Herb. D.-S.; Great Pool, Grose.

E. montanum L.   Broad-leaved Willowherb.   Native.
On and by walls.   Very rare.
M.A.-.-.-. : -.-.-.-.-.   First record: Lousley, 1967.
St Mary's: Wall, Porth Cressa, 1952, Grose. St Agnes: beside
path at foot of Lighthouse wall, 1948, Wanstall and Ribbons; near
Lighthouse, 1952, Grose.

E. tetragonum L.   Native.   Gardens and bulbfields.
Rare.   M.-.T.-.-. : -.-.-.-.-.   First record: Lousley, 1967.
subsp. *tetragonum* (*E. adnatum* Griseb.). St Mary's: Holy Vale;
garden of bungalow by Cinema, Hugh Town, Mrs O. Moyse
(763—det. G. M. Ash). Tresco: Abbey Grounds.
subsp. *lamyi* (F. W. Schultz) Nyman (Tresco: bulbfield, New
Grimsby (505—det. G. M. Ash). I am not satisfied that this subspecies can be satisfactorily distinguished from the last.

E. obscurum Schreb.   Native.
Marshes, cliffs and weed in cultivated ground.   Local.
M.-.T.-.-. : -.-.-.-.-.   First record: Townsend, 1864.
St Mary's: In St Mary's Marshes, Townsend l.c. and Herb. Towns-
end; Higher Moors (266a *fide* Ash); cliff near Porthellick (495,
det. Ash); Old Town Churchyard (920). Tresco: weed in Abbey
Garden (721, det. Ash).

E. palustre L.   Marsh Willowherb.   Native.
Marshes.   Very local.
M.-.-.-.-. : -.-.-.-.-.   First record: Ralfs, 1879.
St Mary's: Higher Marsh, Ralfs' *Flora*: Higher Moors (266, *fide*
Ash); Lower Moors, abundant (758).

Oenothera L.
O. erythrosepala Borbás   Evening Primrose.   Established
alien.
Sand-dunes and waste places.   Local.
M.-.T.-.-. : -.-.-.-.-.   First record: Lousley, 1967.

St Mary's: 1966, P. Z. MacKenzie. Tresco: Abundant in and around the old tumbledown greenhouses between Pentle Bay and Great Pool; fields, Old Grimsby, 1967, F. Russell Gomm.

Fuchsia L.
F. magellanica Lam. Bryher: one bush on rough ground near The Town, 1963, not obviously planted.   Alien.

Circaea L.
C. lutetiana L.   Common Enchanter's Nightshade.   Native?
Cottage wall and waste ground.   Very rare.
M.-.-.-.-. : -.-.-.-.-.   First record: Salmon in Davey, 1909.
St Mary's: Cottage wall, C. E. Salmon; Star Castle Hill, 1952, Grose; garden of Benham's, Garrison, 1970.

HALORAGACEAE
Myriophyllum L.
M. alterniflorum DC.   Alternate-flowered Water-milfoil.
Native.
In large pools.   Very local.
-.-.T.-.-. : -.-.-.-.-.   First record: Townsend, 1864.
Tresco: Tresco Lake, Townsend; Big Pool, Herb. D.-S.; Great Pool; Abbey Pool.

CALLITRICHACEAE
Callitriche L.
C. stagnalis Scop.   Water Starwort.   Native.
Ponds, on mud by pools etc., damp corners of bulbfields.
Occasional.   M.-.T.-.-. : S.-.H.-.-.
First record: Townsend, 1864 as *C. platycarpa* Kütz.
St Mary's: Hugh Town Marsh, Townsend; pond in Old Town Marsh (403); Higher Moors (411a—det. Schotsman); Newford (134a—det. Schotsman); Higher Moors (796—dct. Savidge); damp corner of bulbfield near Bant's Carn Battery; dried-up pond near Content. Tresco: 1955 (spec.), Howitt; mud by Abbey Pool.

Fig 9 Common Windbreaks: 1 *Escallonia macrantha*. 2 *Hebe lewisii*. 3 *Pittosporum crassifolium* (fruit x$\frac{1}{2}$). 4 *Euonymus japonicus* (flower x2$\frac{1}{2}$). 5 *Brachyglottis repanda* (flower x3)

2

3

5

Samson: (486a). St Helen's: by ruins. Probably more common than these records indicate but when growing with *C. obtusangula* it is easily overlooked.

C. obtusangula Le Gall   Native.
In ponds.   Common on St Mary's.
M.-.-.-.-. : S.-.-.-.-.   First record: Tellam in *Bot. Rec. Club. Rep.*, **1877.**
St Mary's: 'The common species at St Mary's (*fide* Mr Briggs and Mr Hanbury)' Ralfs ex Marquand, 1893; 'The common form and probably the *C. verna* of your flora', Ralfs in litt. to Townsend 1877; Newford (134); pond by Watermills Lane (409 and 678); pond in Old Town Marsh (402); Higher Moors (411 teste Schotsman). Samson: pool on south-west shore (486). This species is most frequent in brackish waters and no doubt occurs also on other islands.
C. platycarpa Kutz.   Townsend, 1864 records this as common under the name *C. verna* L. but, as Ralfs pointed out, he confused it with the preceding species.

C. intermedia Hoffm.   Native.
In pools and mud-splashes.   Common.
M.-.-.B.MN. : -.-.-.-.-.   First record: Townsend, 1864 as *C. hamulata* L. var. *pedunculata* DC.
St Mary's: Near Maypole (138); Newford (136, small state); roadside pool between Normandy and Pelistry (679); Watermill Lane (678). Bryher: Water-meadow Pool (446); mud form in puddle on track from The Town to Pool (135). St Martin's: Small pools above Middle Town (as var. *pedunculata*), Townsend; the only form seen in four pools examined (var. *pedunculata*), Reid in Davey, 1909; School Green (427).

ARALIACEAE
Hedera L.
H. helix L.   Ivy   Native.
Walls and rocks.   Common.
M.A.T.B.MN. : -.-.H.TN.E.   First record: Millett, 1852.

UMBELLIFERAE
Hydrocotyle L.
H. vulgaris L.   Marsh Pennywort.   Native.   Common.

Marshy fields, ditchsides, margins of pools, and on sandy tracks.
M.A.T.B.MN. : -.-.-.-.-.   First record: Millett, 1852.

Eryngium L.
E. maritimum L.   Sea Holly.   Native.
Sandy shores.   Frequent.
M.A.T.B.MN. : -.-.-.TN.-.   First record: Heath, 1750.

Anthriscus Pers.
A. caucalis Bieb. (*A. vulgaris* Pers., non Bernh.).   Bur Chervil.
Native.
Sandy places including bulbfields and roadsides.   Local.
-.-.T.-.MN. : -.-.-.-.-.   First record: Lousley, 1939.
Tresco: weed in bulbfields, Dolphinstown, 1957; cliff edge, Old
Grimsby, M. B. Gerrans. St Martin's: Middle Town; abundant
and fine, and especially in bulbfields on the south side of the island.
When I first found this species on St Martin's it was in small
quantity and the great increase in the last thirty years suggests
that it is a recent arrival.

A. sylvestris (L.) Hoffm.   Cow Parsley.   Native.
Roadsides and waste places.   Locally plentiful.
M.-.-.-.-. : S.-.-.-.-.   First record: Ralfs in Marquand, 1893.
St Mary's: Ralfs, 1893; frequent in the Hugh Town, Old Town,
Porthloo area; Salakee Lane; Trenoweth; Watermill Bay, etc.
Samson: about the ruins.
The absence of this species from four of the inhabited islands suggests
that it may be a fairly recent arrival. On Samson it was probably
introduced before the houses were abandoned in 1855.
A. cerefolium (L.) Hoffm. Garden Chervil. Recorded from the vegetable
garden at Tresco Abbey by Miss Anne Dorrien-Smith in Thurston and
Vigurs, 1927. No doubt cultivated.

Scandix L.
S. pecten-veneris L.   Shepherd's Needle.   Colonist.
Cultivated fields and roadsides.   Rare.
M.-.-.-.MN. : -.-.-.-.-.   First record: Lousley, 1939.
St Mary's: Field beyond Parting Carn, Herb. D.-S.; field near
Parting Carn, 1940 and 1959; Between Hugh Town and Old Town
(131); abundant in one field on Penold Farm, 1967, P. Z. Mac-
Kenzie. St Martin's: bulbfield by road between Lower and Middle
Town, 1967.

Myrrhis Mill.

M. odorata (L.) Scop. Sweet Cicely. This was recorded by Miss Anne Dorrien-Smith from Tresco in Thurston and Vigurs, 1927. In 1939 she confirmed that this had been planted in the orchard at Tresco Abbey.

Torilis Adans.

T. japonica (Houtt.) DC. Hedge Parsley. Recorded as *Caucalis anthriscus* Huds. from Hugh Town, St Mary's in Smith, 1909. This species is common in Cornwall, but its claim to occur in Scilly rests on this solitary record, which may be an error. Confusion with *Anthriscus sylvestris*, which is common about Hugh Town, is suspected.

T. nodosa (L.) Gaertn.   Knotted Hedge Parsley.   Native.
Roadsides and dry places and sandy cultivated fields.   Frequent.
M.A.T.B.MN. : -.-.-.-.-.   First record: Townsend, 1864.
St Mary's: Maypole, Grose; Old Town. St Agnes: roadside on ascent from Gugh, Herb. Dallas. Tresco: Ralfs' *Flora*; Bryher: Southward, Grose; The Town. St Martin's: Higher Town, Dallas; Middle Town.

Smyrnium L.

S. olusatrum L.   Alexanders.   Naturalised alien.
Waste places and roadsides.   Common.
M.A.T.B.MN. : -.-.-.-.-.   First record: Millett, 1852.
An ancient introduction as a potherb, now completely naturalised. 'Horses and donkeys prefer the wild and prolific weed Alexanders to grass or better food', *Scillonian*, **20,** 174, 1946.

Conium L.

C. maculatum L.   Hemlock.   Native.
M.A.T.B.MN. : S.-.H.-.-.   First record: Millett, 1852.
Waste places and about buildings.   Frequent.
Probably introduced on Samson and St Helen's.

Bupleurum L.

B. lancifolium Hornem.   Casual.
Weed in a garden on The Garrison, St Mary's, 1967, P. Z. Mac-Kenzie. Also as a weed in the garden of 'Justholm', Old Grimsby, Tresco, 1968, Mrs D. Finucane. Probably introduced with cage-bird seed.

Apium L.

A. graveolens L.   Wild Celery.
Recorded by Townsend, 1864 from 'Shores in several places', but there is no specimen in his herbarium. It is also included in the list of Boyden, 1893, and marked as seen in the period 1903 to 1909 in Davey, 1909. Although common in Cornwall, there are few suitable habitats in Scilly, and the records cannot be accepted in the absence of a specimen.

A. nodiflorum (L.) Lag.   Procumbent Marshwort.   Native.
Ditches and margins of pools.   Local.
M.A.T.-.-. : -.-.-.-.   First record: Townsend, 1864.
St Mary's: Higher and Lower Moors; Watermill and Pungies Lanes; Salakee. St Agnes: Ralfs' *Flora*. Tresco: east end of Big Pool, Herb. D.-S.; By a small pool adjoining Pool Road.
var. *ochreatum* DC. is recorded by Boyden in Davey, 1909.

A. inundatum (L.) Reichb.f.   Least Marshwort.   Native.
Freshwater pools.   Rare.
M.-.T.B.-. : -.-.-.-.   First record: Townsend, 1864.
St Mary's: Ralfs' *Flora*. Tresco: Tresco Lake, Towns. Bryher: Freshwater pool, Towns.; in small pools adjoining The Pool.

Petroselinum Hill
P. crispum (Mill.) Airy Shaw   Parsley.   Established alien.
Walls and cliffs.   Rare.   M.-.T.-.MN. : -.-.-.-.
First record: Smith, 1909, but probably found earlier by Boyden (see Davey, 1909) though it is not included in his 1893 list.
St Mary's: Hugh Town, Smith; plentiful about Woolpack Point on The Garrison (130 and 255) 1938, increasing and spread to near coastguard houses 1963. Tresco: Smith, 1909. St Martin's: on wall Upper Town, 1956, established there 1963.

Conopodium Koch
C. majus (Gouan) Loret   Pignut.   Native.
Rough pastures and cliff slopes, usually associated with bracken. Rare.   M.-.T.-.MN. : -.-.-.-.   First record: Townsend, 1864,
St Mary's: Ralfs' *Flora*. Tresco: Dial Rocks, Herb. D.-S.; abundant under bracken in Gimble Porth below Beacon Hill. St Martin's: 1 specimen near Higher Town, Townsend; under brambles by path from Higher Town to Daymark, Dallas; in quantity under wall between Higher Town and Culver Hole.

Pimpinella L.

P. saxifraga L.   Burnet Saxifrage.   Native.
By cliff path on blown sand over granite.   Very rare.
M.-.-.-.-. : -.-.-.-.-.   First record: Somerville, 1893.
St Mary's: cliff north-east of Lifeboat Station, Grose, 1952;
between Porth Mellin and Thomas' Porth, 1952, 3 plants 1956.

Aegopodium L.

A. podagraria L.   Goutweed.   Denizen.
Weed about houses and gardens.   Very rare.
M.A.-.-.-. : -.-.-.-.-.   First record: Lousley, 1967.
St Mary's: M. Knox, 1938; under hedge in Star Castle garden,
1940; garden in Hugh Town, 1970, K. E. Bull. St Agnes: 1960,
E. G. Philp.

Berula Koch

B. erecta (Huds.) Coville   Lesser Water Dropwort.   Native.
Ditches and marshes.   Local.   M.-.-.-.-. : -.-.-.-.-.
First record: Lawson, 1870, without locality.
St Mary's: Townsend in Ralfs' *Flora;* north side Porthellick Pool;
common in Lower Moors.

Crithmum L.

C. maritimum L.   Rock Samphire.   Native.
Maritime rocks, thriving under heavy concentrations of salt.
Abundant.
M.A.T.B.MN. : S.AT.H.TN.E.   First record: Heath, 1750.
The abundance of Rock Samphire is still an important botanical
feature and was mentioned by all the early writers. Robert Heath,
in his description of Tresco, wrote in 1750 'Samphire, of an extraordinary Kind, is produced here, and in other of the Off-islands,
in Abundance, and is used both for Distilling and Pickling. The
Method of Preserving it for Pickling, at any Time, is, by putting
it into small Casks, and covering it with a strong Brine of Salt and
Water, which changes it yellow; but Vinegar restores its Green-
ness in Pickling. Being preserved after this Manner, it is sent in
small Casks to distant Ports for Presents.'. He also says it was
gathered on Bryher. Borlase, in 1756 says 'Sampier they have of
the best and largest kind (far superior to the Cornish)', and
Woodley, in 1822, likewise praises the quality and abundance of
Samphire and refers to its export in casks.

Oenanthe L.
O. fistulosa L.   Tubular Water Dropwort.   Native.
Marshes and margins of pools.   Locally plentiful.
M.-.T.-.-. : -.-.-.-.-.   First record: Townsend, 1864.
St Mary's: Common in Hugh Town and Old Town Marshes,
Townsend; still at Higher Moors, and in part of Lower Moors
near Rocky Hill. Tresco: west end of pool, Herb. D.-S.; Great
Pool, Grose; reed-bed near Abbey; by Pool Road.
This species is very rare in Cornwall, and rare in Devon.

O. lachenalii C. C. Gmel.   Parsley Water Dropwort.   Native.
Marshy ground by lake.   Very rare.   -.-.T.-.-. : -.-.-.-.-.
Only record: By the long lake behind Tresco Abbey, July 21, 1923,
F. J. Hanbury (Herb. Mus. Brit.).

O. crocata L.   Hemlock Water Dropwort.   Native.
Marshes, pool, ditches and (rarely) wall tops.   Locally common.
M.-.-.B.-. : -.-.-.-.-.   First record: Townsend, 1864.
St Mary's: Common in Hugh Town and Old Town Marshes,
Townsend; still common in Higher and Lower Moors; Watermill
Cove; roadside on the Garrison. Bryher: Wall-top near Meadow
Pool.
O. crocata is able to grow in Scilly in very much drier places than
is usual on the mainland.

Aethusa L.
A. cynapium L.   Fool's Parsley.   Casual.
Very rare.
M.-.-.-.-. : -.-.-.-.-.   First record: Townsend, 1864.
References by Ralfs, 1879 and Davey, 1909 seem to be based on
Townsend who found a single plant near Hugh Town.

Foeniculum Mill.
F. vulgare Mill.   Fennel.   Native.
Roadsides and waste places, mainly near the shore.   Common.
M.A.T.B.MN. : -.-.-.-.-.   First record: Millett, 1852.
Abundant about Hugh Town and Old Town, St Mary's, and in
St Agnes and Bryher, and less plentiful on Tresco and St Martin's.
Withered stems of Fennel are often used for lighting fires (*Scillonian*,
20, 19, 1946).

Angelica L.

A. sylvestris L.   Wild Angelica.   Native.

Marshes and other wet places.   Local. (Map 9, p 192).

M.-.-.-.MN. : -.-.-.-.E.   First record: Townsend, 1864.

St Mary's: Very luxuriant in marshes, attaining 9 ft, Smith, 1912; Higher and Lower Moors; Green, Grose. St Martin's: 1862, Herb. Townsend; near Turfy Hill. Great Ganinick: north end (327).

Pastinaca L.

P. sativa L.   Parsnip.   Native.

Roadsides, waste places, etc.   Rare.

M.-.-.-.MN. : -.-.-.-.   First record: Townsend, 1864.

St Mary's: west of Hugh Town, Townsend; Star Castle Hill, Grose; Garrison Wall. St Martin's: near Higher Town, Townsend; St Martin's Bay, Grose.

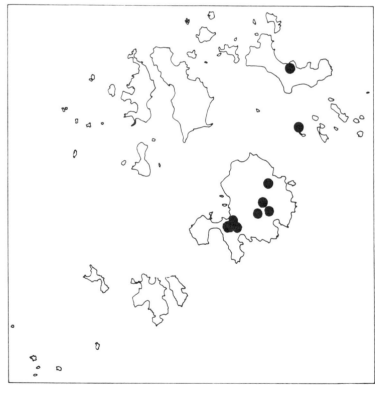

Angelica sylvestris, Wild Angelica

TS CHARACTERISTIC OF ROCKS AND WALLS: (*top*) Wall Pennywort, *Umbilicus rupestris;* (*bottom*) Sea Spleenwort, *Asplenium marinum*

Heracleum L.

H. sphondylium L.   Hogweed.   Native.

Roadsides, field borders, rough cliff slopes.   Very common.

M.A.T.B.MN. : S.-.H.TN.E.   First record: Millett, 1852.

Daucus L.

D. carota L.   Wild Carrot.   Native.

Roadsides, sandy ground, cultivated fields, etc.   Common.

M.A.T.B.MN. : -.-.-.-.E.   First record: Millett, 1852.

On uninhabited islands recorded only from Nor-nour in the Eastern Isles, by Grose.

The common Carrot in Scilly is subsp. *carota*, but on headlands close to the sea plants occur which seem indistinguishable from subsp. *gummifer* Hook.f. as this occurs in Cornwall. For example, specimens collected from Woolpack Point, St Mary's in 1939 (256) closely resemble plants I gathered from Gurnard's Head, Zennor in 1951 as being at or near the *locus classicus* for *D. maritimus* With. Although some of the characters are retained in cultivation, I am unable to satisfy myself that hispid rays occur constantly in umbels flat in fruit, or with leaves narrower in outline with fleshy leaflets. Townsend, 1864 included '*D. gummifer* Lam.' as common but there is no supporting specimen in his herbarium. I regard all Wild Carrots in Scilly as variations of a single taxon.

EUPHORBIACEAE

Mercurialis L.

M. annua L.   Annual Mercury.   Colonist.

Roadsides, cultivated ground and waste places.   Local.

M.A.T.B.MN. : -.-.-.-.   First record: Townsend, 1864.

St Mary's: common about Hugh Town, Townsend; still common there and elsewhere in the island. St Agnes: Middle Town. Tresco: occasional. Bryher: Northward; Pool. St Martin's: Higher Town. frequent; Middle Town.

Townsend's record was also the first for Cornwall, where it is rare and regarded as an alien. In Scilly it is very much a 'follower of Man' and an introduction. Occasionally female plants occur with narrow, almost linear, leaves, as at Old Town, St Mary's, 1952 (501).

Euphorbia L.

E. peplis L.   Purple Spurge.   Native.

Just above high-water mark on fine shingle shores.   Very rare.
M.A.-.-.-. : -.-.-.-.   First record: Joseph Woods, 1852.
St Mary's: Porthellick Bay, 1900, Milne in Davey, *J. Roy. Inst.
Cornw.*, **50**, 20 and 1905, Salmon in Davey, 1909. St Agnes:
August 17, 1852, Woods in Herb. S.L.B.I., and *Rep. Penzance Nat.
Hist. Soc.*, **1852**, 100.; 'I have searched for it in vain', Ralfs ex
Marquand, 1893; 15 plants, White in *J. Bot.*, **52**, 19, 1914.
Purple Spurge, a drift-line species, is erratic in appearance in all
its localities. Even in Alderney it fluctuates widely in numbers from
year to year. It is probably dependent on the incidence of storms
of the previous winter and if these shift tons of shingle over its
habitat there may be a long interval before the seeds of this annual
spurge are again able to germinate and grow. The exact date of the
last appearance in Scilly is not known, but A. J. Hosking sent a
specimen from Scilly to Downes in 1920, and Thurston stated that
it still existed in 1936. Suitable habitats suffer very much more
from trampling than they did earlier, but the species is likely to
reappear. (The date 1950 quoted in Lousley, 1967 was based on the
B.S.B.I. Maps Scheme and is incorrect.)
E. platyphyllos L. Error. Recorded by Boyden, 1890, but Davey, 1909
says that Arthur Bennett saw the specimen and determined it as *E.
helioscopia*.

E. helioscopia L.   Sun Spurge.   Colonist.
Weed in cultivated fields and gardens.   Not common.
M.A.T.B.MN. : -.-.-.-.-.   First record: Townsend, 1864.

E. peplus L.   Petty Spurge.   Colonist.
Weed in gardens.   Rare.
M.A.T.B.MN. : -.-.-.-.-.   First record: Townsend, 1864.
St Mary's: Hugh Town, Grose; garden of Tregarthen's Hotel;
Garden at Maypole; under wall at Parting Carn. St Agnes: Middle
Town, Grose. Tresco: Abbey Gardens; gardens at Dolphin's
Town; garden of New Inn. Bryher: Southward. St Martin's: seen
twice.
In Scilly this is almost restricted to gardens and I have never seen
it in bulbfields or other arable.

E. portlandica L.   Portland Spurge.   Native.
Sandy shores and dunes, and rough cliff slopes.   Frequent.
M.A.T.B.MN. : S.-.-.TN.E.   First record: Bree in Forbes, 1821.

E. paralias L.   Sea Spurge.   Native.
Sandy shores, and dunes.   Common.
M.A.T.B.MN. : S.-.-.TN.E.   First record: Bree in Forbes, 1821.
These two species sometimes grow together in quantity.

E. amygdaloides L.   Wood Spurge.   Native.
Rough places near the coast.   Local. (Map 10, p 196.)
-.-.T.B.MN. : -.-.H.TN.E.   First record: Townsend, 1864.
Tresco: Bushy ground near Cromwell's Castle, Townsend; Monument Hill; Carn Near; Gimble Porth. Bryher: west side of Samson Hill. St Martin's: near shore below the old school. St Helen's: south side. Tean: Dallas. Eastern Isles: Great Ganinick (329); Great Ganilly. Absent from the southern part of the archipelago—St Mary's, Samson and St Agnes.

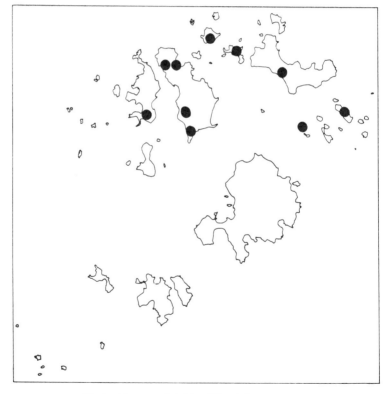

Euphorbia amygdaloides, Wood Spurge

POLYGONACEAE

Polygonum L.

P. aviculare L. *sensu lato.*   Knotgrass.   Native.
Cultivated land, roadsides and waste places.   Very common.
M.A.T.B.MN. : -.-.-.-.-.   First record: Millett, 1852.
The aggregate species is now divided into segregates, of which the
following occur:—

P. aviculare L. (*P. heterophyllum* Lindm.) First record: Lousley,
1940. This is common on all the inhabited islands in cultivated
fields, and also roadsides and waste places. It sometimes occurs
with fleshy leaves on beaches (Porthcressa (379)). Var. *agrestinum*
(Jord) reported by Tellam in Davey, 1909 probably belongs here.

P. arenastrum Bor. (*P. aequale* Lindm.; *P. littorale* auct.). First
record: Lousley, 1940. This also occurs on all the inhabited islands
but is much less common and mainly in sandy places. Here belong
specimens from St Mary's as 'var. *littorale* Koch' gathered in 1873
by W. H. Beeby (Herb. S.L.B.I.).

P. raii Bab.   Ray's Knotgrass.   Native.
Sandy and fine shingle shores.   Very rare.
M.-.-.-.MN. : -.-.-.-.-.   First record: Ralfs, 1879, First evidence:
Beeby, 1873—see below.
St Mary's: Near Old Town, 1873, Beeby in Herb. S.L.B.I.; near
Old Town and Porthellick, C. E. Salmon in Davey, 1909. St
Martin's: 1876, Ralfs' *Flora*.

P. maritimum L.   Sea Knotgrass.   Native.
Sandy shores.   Very rare and erratic.
M.-.T.-.-. : -.-.-.-.-.   First record: Curnow, 1878.   First evidence: Woods, 1852.
St Mary's: Hugh Town, Joseph Woods, 1852 in Herb. S.L.B.I.;
Crow Sound on 'compact sandy flat', 1878, Curnow, Herb. Mus.
Brit. and *Bot. Soc. Rec. Club Rep.*, **1878,** 11. Tresco: north of Tresco,
Smith, 1909.
This species, like the last, is readily recognised when fully mature
but young material can be deceptive. I have no doubt that Woods'
specimen is correct, but Curnow's is a poor one, and I have not
seen material collected by Smith. Both *P. maritimum* and *P. raii*,

like *Euphorbia peplis* and other strand plants, fluctuate widely in numbers and appearance from year to year, and the fact that they have not been seen recently is no reason why they should not reappear.

P. amphibium L.   Amphibious Bistort.   Native?
In a pond.   Very rare.
-.-.T.-.-. : -.-.-.-.-.   First record: this *Flora*.
Tresco: Abbey Pool, north side, 1968, R. Lancaster. This species is native in Cornwall and probably freely distributed from pond to pond by ducks. As it is also commonly grown in ornamental ponds the status must be regarded as doubtful.

P. persicaria L.   Common Persicaria.   Native.
Cultivated fields, roadsides and near pools.   Common.
M.-.T.B.MN. : -.-.-.-.-.   First record: Millett, 1852.

P. lapathifolium L. (*P. nodosum* Pers.).   Pale Persicaria.
Native.
Cultivated ground, marshes, waste places.   Rather local.
M.-.T.-.-. : -.-.-.-.-.   First record: Townsend, 1864.
St Mary's: Old Town Marshes, Maypole, Holy Vale, etc.
Tresco: Pool Road and near Tresco Abbey Gardens.

P. nodosum was listed by Townsend in addition to *P. lapathifolium*, and the late C. E. Britton named material of both for me. Recent research has shown that they cannot be kept up as separate species and it is significant that Britton named both from the same localities in St Mary's and Tresco.

P. hydropiper L.   Waterpepper.   Native.
Pondsides, damp tracks and moist hollows.   Rather rare.
M.-.T.-.-. : -.-.-.-.-.   First record: Lawson, 1870.
St Mary's: Lawson, 1870; Ralfs' *Flora;* track near Bar Point; roadside pool, Maypole; wet lane near Content; ponds by roadside, Lower Newford. Tresco: track near Abbey Pool; near Great Pool; hollow on Castle Down.

P. convolvulus L.   Black Bindweed.   Native.
Cultivated and waste ground.   Rather common.
M.A.T.B.MN. : -.-.-.-.-.   First record: Millett, 1852.

P. cuspidatum Sieb. and Zucc.   Japanese Knotweed.   Established alien.

Roadsides.   Rare.

M.A.T.-.-. : -.-.-.-.-.   First record: Lousley, 1940.

St Mary's: Hugh Town and Holy Vale, 1939, Grose; roadside, Rose Hill, 1939, 1956 (754). St Agnes: Field side, 1953, Polunin. Tresco: corner of field, Back Lane, 1968, D. A. Cadbury.

P. aubertii L. Henry (*P. baldschuanicum* Regel). Noted in 1970 as rampant along a lane in Bryher but only just out of a garden.

Fagopyrum Mill.

F. esculentum Moench   Buckwheat.   Casual.

Found by Grose in 1952 near a chicken-run between Bar Point and Innisidgen Carn, St Mary's.

Muehlenbeckia Meisn.

M. complexa (Cunn.) Meisn.   Naturalised alien.

Walls, rocky knolls, quarries.   Locally abundant. (Fig 6, p 95.)

M.A.T.-.-. : -.-.-.-.-.   First record: Smith, 1909 (without locality).

St Mary's: about Hugh Town on walls; rocks on Buzza Hill; hill above the Church; copiously along hedges in Pungies Lane; quarry between Hugh Town and The Telegraph. St Agnes: near Parsonage. Tresco: walls near Abbey grounds (237); established on Appletree Banks.

This small New Zealand shrub was established in Herm, in the Channel Isles, as early as 1911 (Herb. Kew and *Bot. Mag.*, t. 8449). and is still abundant there, but Smith reported it as established in Scilly two years earlier. It spreads freely and the great mass on the south-west side of Buzza Hill increased rapidly between 1939 and 1952, and has continued to spread since. Here, and elsewhere, in Scilly it is an impressive sight, smothering all other vegetation with its relentless spread.

Rumex L.

R. acetosella L.   Sheeps' Sorrel.   Native.

Stone walls, sandy cultivated and waste ground, heaths. Abundant.

M.A.T.B.MN. : S.AT.H.TN.E.   First record: Townsend, 1864.

A variable species.

R. acetosa L.   Common Sorrel.   Native.

Cliff slopes, dunes, roadsides, walls, etc.—almost ubiquitous. Abundant.

M.A.T.B.MN. : S.-.H.TN.E.   First record: Millett, 1852.

Variable. An interesting variant with fruiting valves measuring only 3 x 3 mm occurs in the Eastern Isles: Great Ganilly (317); Great Arthur (316)—see *J. Bot.*, **78**, 156, 1940.

R. hydrolapathum Huds.   Great Water Dock.   Native.

In a marsh.   Rare.

M.-.-.-.-. : -.-.-.-.-.   First record: Townsend, 1864.

St Mary's: Marshes, Townsend; Old Town Marsh, Ralfs, 1877a. I have failed to refind this after careful search, and the species is probably extinct.

x R. obtusifolius=R. x weberi Fisch-Benz. In 1877 T. R. A. Briggs, who knew this hybrid at Downderry, E. Cornwall, was on holiday in Scilly and in crossing Old Town Marsh he drew Ralfs' attention to plants he believed to be this. Immature material was collected by Briggs, Tellam and Curnow in 1877 and 1878, and a root was sent to Arthur Bennett to grow at Croydon. From this he distributed immature material in 1880 and 1881 and radical leaves in 1883. These gatherings are represented in many public herbaria, and one in Herb. Kew, collected by Curnow in July 1878, has been determined by Rechinger in 1933 as *R. hydrolapathum x obtusifolius* subsp. *agrestis*. It certainly resembles this hybrid, which I know well from elsewhere, but in the absence of fruit doubt must remain.

R. crispus L.   Curled Dock.   Native.

Cultivated fields, roadsides, dunes and shores.   Very common.

M.A.T.B.MN. : S.AT.H.TN.E.   First record: Townsend, 1864.

A species showing a wide range of variation. On shores the leaves are often fleshy and there are large swollen tubercles on each of the three tepals (var. *littoreus* Hardy).

x R. obtusifolius=R. x acutus L. This hybrid forms freely whenever the parents grow fairly close to one another. I have noted it as common on St Mary's and St Agnes, very common on Bryher, and on Tresco at Old Grimsby and the Duckery shore—see Lousley, 1940.

x. R. pulcher=R. x pseudopulcher Hausskn. A plant believed to be of this parentage was seen in Abbey Gardens, Tresco in 1939.

R. obtusifolius L.   Broadleaved Dock.   Native.
Roadsides, field borders and waste places.   Very common.
M.A.T.B.MN. : -.-.-.-.-.   First record: Townsend, 1864.
In contrast to *R. crispus*, Broadleaved Dock is always associated
with human activities and this is reflected in its absence from the
uninhabited islands. Only subsp. *obtusifolius*, which is native,
occurs in Scilly.
x R. pulcher = *R. x ogulinensis* Borbás. This very rare hybrid grows
with the parents by cottages at Middle Town, St Martin's (425).

R. pulcher L.   Fiddle Dock.   Native.
Roadsides, pastures, cultivated land and waste places, especially in
dry and sunny spots.   Abundant.
M.A.T.B.MN. : -.-.-.TN.-.   First record: Townsend, 1864.
R. pulcher, which has its main centre round the Mediterranean,
is exceptionally abundant in Scilly. It is sometimes not recognised
by summer visitors because the fiddle-shaped outline of the leaves
is not well marked at that time. The frequency of the species in
bulbfields suggests that the seeds germinate here in autumn, and
reach maturity and flower the following summer.
x R. rupestris = *R. x trimenii* Camus This exceedingly rare hybrid
has been found twice:— Tresco: New Grimsby (R.B.E. 270) and
Samson: east coast (R.B.E. 272). On Samson the hybrid occurred
with abundant *R. rupestris* but over half a mile of sea separated it
from the nearest *R. pulcher* on Bryher and nearly a mile from the
nearest on Tresco. This is an excellent example of the transport of
Rumex pollen by wind.

R. sanguineus L.   Wood Dock.   Native.
Damp shady places.   Very rare.
M.-.T.-.-. : -.-.-.-.-.   First record: Townsend, 1864.
St Mary's: In north of island, 1939. Tresco: Abbey Grounds, 1939;
near Abbey, 1956; damp ride by Great Pool, 1957.

R. conglomeratus Murr.   Clustered Dock.   Native.
Marshes and roadsides.   Frequent.   M.A.T.B.-. : -.-.-.-.-.
First records: Boyden, 1893 and Somerville, 1893.
Although characteristic of wet places this is by no means confined
to them.
x R. pulcher = *R. x muretii* Hausskn. Bryher: Southward (340),
J. Bot., **78**, 156, 1940.

R. rupestris Le Gall   Shore Dock.   Native.
Sandy and rocky shores.   Common. (Plate p 289).
M.A.T.B.MN. : S.AT.H.-.E.   First record: Beeby in *J. Bot.*,
13, 337, 1875.
St Mary's: Old Town Bay, 1873, W. H. Beeby in Herb. S.L.B.I.
St Agnes: Tellam, 1877, Herb. Tellam; Curnow, 1877, Herb.
Curnow; The Gugh, Ralfs in Marquand, 1893, Salmon, 1905,
Herb. Mus. Brit., still there. Tresco: Reid in Davey, 1909; near
Carn Near. Bryher: Pool Bay and in two coves on south-east coast.
St Martin's: Reid in Davey, 1909; White Island (381). Samson:
common. Annet: in several places. St Helen's: scarce. Eastern
Isles: Great or Middle Arthur, Herb. Dallas; Nor-nour (326).
R. rupestris occurs only on the shores of western Europe and
mainly in north-west France, and is one of the world's rare docks.
On the shores of south-west England and south Wales it is decreas-
ing rapidly as cove after cove gets more overrun by trippers, and
the population in Scilly is therefore of special importance.
R. triangulivalvis (Danser) Reching.f. Error. The record in Lousley,
1967 was based on the assumption that a poor specimen in Herb.
Dorrien-Smith was part of the same gathering as a specimen in Herb.
Kew collected by Miss Gwen Dorrien-Smith. The latter, as I now
know, was found by her in North America in 1926.

URTICACEAE

Parietaria L.
   P. judaica L. (*P. diffusa* auct.)   Pellitory-of-the-Wall.   Native.
   On stone walls.   Frequent.
   M.A.T.B.MN. : -.-.-.-.-.   First record: Millett, 1852.

Soleirolia Gaud.
   S. soleirolii (Req.) Dandy (*Helxine soleirolii* Req.)   Mind-your-
   own-business.   Established alien.
   Gutters, the base of walls, and other damp shady places.   Local.
   M.A.T.-.-. : -.-.-.-.-.   First record: Lousley, 1967.
   St Mary's: near Garrison Gateway, 1939; Garrison Hill, 1953 (583);
   in gutter, Old Town, 1952; several places about Hugh Town and
   Old Town, 1963; London, 1956. St Agnes: near Lighthouse, 1952,
   Grose; wall near Parsonage, 1953. Tresco: New Grimsby and
   established about Abbey Gardens.

Urtica L.
U. urens L.   Small Nettle.   Native.
Cultivated fields and gardens and places rich in nitrogen.
Common.
M.A.T.B.MN. : S.-.-.-.E.   First record: Townsend, 1864.
On Samson and Great Innisvouls this species is found on ground
very rich with the droppings of sea-birds.

U. dioica L.   Common Nettle.   Native.
Waste ground, cultivated fields and roadsides.   Common.
M.A.T.B.MN. : S.AT.H.TN.E.   First record: Millett, 1852.
On the uninhabited islands this species occurs mainly where there
is a lot of sea-bird guano, or by ruins as on Samson, St Helen's
and Tean.

CANNABACEAE

Humulus L.
H. lupulus L.   Hop   Native?
Hedges.   Rare.
M.-.T.-.-. : -.-.-.-.-.   First record: Millett, 1852.
St. Mary's Ralfs ex Marquand, 1893; Watermill Bay; Hugh Town;
Green, Grose. Tresco: Back Lane.
Hop is a strong-growing perennial able to persist under adverse
conditions and some or all of these localities may be due to former
cultivation.

ULMACEAE

Ulmus L.
Traces of fossil elm pollen have been found at Innisidgen by
Prof G. W. Dimbleby, but all the elms now growing in Scilly are
believed to have originated from plantings for shelter purposes.
It is known that elms were planted at Holy Vale about 1650, and
by 1695 there was one tree with a trunk over three feet in circumference (Turner in *Scillonian*, **159**, 154, 1964). A century later
there were very fine trees there (Woodley, 1822 p 220). Borlase in
1756 had recommended the planting of Dutch Elm (cf. North, 1850
p 91) but others were also used. Now elms planted as windbreaks
are common, and occur on all the inhabited islands, but their
identification presents difficult problems. Many are dwarfed and
windswept, or mutilated and replaced by sucker growths, or

otherwise untypical. I am indebted to Dr R. Melville for naming some of my specimens but the following account may need modification.

U. glabra Huds.   Wych Elm.   Established alien.
Roadsides, hedges.   Uncommon.
M.-.T.-.-. : -.-.-.-.-.   First record: Lousley, 1967.
St Mary's: Holy Vale. Tresco: Pool Road; Old Grimsby.
Specimens collected on Garrison Hill, St Mary's (649) were determined by Dr Melville as a hybrid of this species but too 'juvenile' for more exact determination.

U. procera Salisb.   Common Elm.   Established alien.
Roadsides and hedges.   Frequent.
M.A.T.B.MN. : -.-.-.-.-.   First record: Townsend, 1864.
Specimens from Love Lane, St Agnes, 1954 (686) were named by Dr Melville as 'the suberous form which is not uncommon in the west country'.

U. x hollandica Mill.   Dutch Elm.   Established alien.
Roadsides, thickets, hedges.   Common.
M.A.T.-.MN. : -.-.-.-.-.   First record: Lousley, 1967.
Specimens have been confirmed for me by Dr Melville from:—
St Mary's: thicket, Tremelethen Marsh, Polunin (710); Old Town Bay (496). I have noted this only from St Mary's, where it is certainly widespread, and the records from the other islands are from R. P. Bowman, who may well be right in regarding this as the most common elm in Scilly.

MORACEAE

Ficus L.
F. carica L.   Fig.   Established alien.
Tresco: near Borough Farm. St Martin's: large old plant in hedge near Lower Town.

BETULACEAE

Betula L.
Although birches are common in Cornwall there is no evidence of their occurrence in Scilly in historic times. Birch pollen has been found by Prof G. W. Dimbleby in peat from below the beach sand at Porthellick, and in deposits at Halangy and Innisidgen.

Alnus Mill.

A. glutinosa (L.) Gaertn.   Alder.   Native.

Margin of lake.   Rare.

-.-.T.-.-. : -.-.-.-.-.   First record: Lousley, 1967.

Tresco: By Great Pool near the Abbey. Fossil pollen has been found by Prof G. W. Dimbleby in deposits at Innisidgen.

CORYLACEAE

Corylus L.

C. avellana L. The absence of Hazel from Scilly in modern times is somewhat puzzling. It is abundant in Cornwall. Hazel nuts have been extracted from the peat beneath beach sand at Porthellick, and the pollen was dominant with oak in a deposit at Innisidgen. (Dimbleby *ined.*). Hazel nuts are also found in the peat of the 'submerged forest' in Mount's Bay.

FAGACEAE

Quercus L.

Q. cerris L.   Turkey Oak.

Planted in the Parsonage garden, St Agnes (815).

Q. ilex L.   Holm Oak.

Planted as a windbreak in several places in Tresco:—above Gimble Porth; Northward; and round Abbey Gardens.

Q. robur L.   Common Oak.   Native.

On site of ancient woodland, and in bramble on small island.

Rare.   -.-.T.-.-. : -.-.-.-.E.   First record: Lousley, 1939.

Tresco: Abbey Grounds; two trees by Abbey Road *c* 100 years old; seedlings at Gimble Porth, A. A. Dorrien-Smith. Eastern Isles: oak about 2 ft 6 in tall in tangle of bracken, bramble and honeysuckle, Great Ganinick 1938, Dallas.

There is ample evidence that the oak was formerly plentiful in Scilly, and perhaps also on the land between the islands now submerged, but was almost eliminated by the inhabitants and shipping to meet their needs for fuel and timber. Pollen in quantity has been found in deposits at Innisidgen, and some in deposits at Halangy, and there were pieces of oakwood in peat below the beach sand at Porthellick (Dimbleby *ined.*). Oak charcoal has been recognised in an Iron Age Site on Nor-Nour (Dimbleby *ined.*), and carbonised wood was found in a cist dating from the first century A.D. on the Old Man of Tean (Tebbutt, *Antiq. J.*, **14**, 302, 1934).

There was a tradition of a wood on the hill north of Tresco Abbey 'which is still called Abbey Wood, and where roots of trees have been dug up, in the memory of man' (Woodley, 1822, 292) and the presence of a wood is necessary to explain certain old charters. In the 'Travels of Cosmo the Third during the reign of King Charles the Second (1669)', which was not published until 1821, we read 'In digging the ground, there are found in many places a great number of very thick stumps of oak, which evidently belonged to trees of extraordinary magnitude'. This is corroborated in a recently discovered manuscript which dates from about 1695. This states that in digging 'bituminous earth' in St Mary's and Tresco 'they ordinarily find Trees lodged against it. He takes them to be Oak and Elm and hath seen of them from 2 foot to 8 in circumference. . . . He never could find any signs of an Ax or Fire upon any of them. . . . He hath seen them lying in the Earth 5 or 6 feet deep.' (Turner, *Scillonian*, **159**, 155, 1964). While there can be no doubt that trees were almost eliminated from the islands before the arrival of Augustus Smith in 1834, it is possible that young oaks too small to be of value persisted on Tresco in addition to those he planted.

SALICACEAE

Populus L.
P. alba L.   White Poplar.   Planted alien.
This grows by Tresco Great Pool in Pool Road, where it was first noticed by J. D. Grose.

P. x canadensis Moench   Canadian Poplar.   Planted alien.
A few trees have been planted on St Mary's: Hugh Town, 1939, Grose; Old Town (310).

P. gileadensis Rouleau.   Balsam Poplar.   Planted alien.
St Mary's: Toll's Island, 1952, Grose, det. N. Y. Sandwith. Still there, but stunted.

Salix L.
S. fragilis L.   Crack Willow.   Established alien.
M.-.-.-.-. : -.-.-.-.-.   First record: Lousley, 1967.
St Mary's: Bar Point, a single very small plant, Grose, 1952; Old Town Marsh, in planted row (924, 925, det. R. D. Meikle), 1970, Lousley and MacKenzie.

Mr MacKenzie tells me that withies are imported from Somerset for the construction of lobster pots and he showed me a row of willows grown from planting the surplus rods. The row is added to from year to year when insufficient withies are left over to make another pot. The species represented are *S. fragilis*, *S. triandra* (see below) and *S. cinerea* ssp. *oleifolia*, which is probably a native intrusion. The plant found by Mr Grose at Bar Point was on the sandy beach and had no doubt grown from a twig washed up by the tide, and is likely to have originated from a lobster pot.

S. triandra L.   Almond Willow.   Established alien.
M.-.-.-.-. : -.-.-.-.-.   First record: this *Flora*.
St Mary's: Old Town Marsh, in planted row (923, 926, det. R. D. Meikle), 1970, Lousley and MacKenzie. See previous entry. This provides a good example of the way *S. fragilis* and *S. triandra* are readily introduced into natural habitats so that much of their distribution in the British Isles is due to planting.

S. viminalis L.   Osier.   Planted alien.
Damp hedges adjoining marshes.   Rare.
M.-.-.-.-. : -.-.-.-.-.   First record: Lousley, 1967.
St Mary's: Old Town, 1939, Lousley and Grose.

S. caprea L.   Sallow.
There is no certain record. R. C. L. Howitt tells me that he thinks he has seen it on St Mary's, and I entered it in my field notebook from a hedge by School Green, Tresco, but took no specimen.
x S. cinerea=*S. x reichardtii* A. Kerner.   St Mary's: Old Town Marsh and Tremelethen, Howitt.

S. cinerea L. subsp. *oleifolia* Macreight (*S. atrocinerea* Brot.) Grey Sallow.   Native
Hedges and about pools.   Locally common.
M.-.T.-.-. : -.-.H.-.E.   First record: Lousley, 1967.
St Mary's: Old Town (174) and all about Lower Moors; The Garrison; Normandy. Tresco: About the Abbey and Great Pool and elsewhere in hedges; Block House Point, Grose. Eastern Isles: Great Arthur, in a small gully, dwarfed (714). St Helen's: 1968, R. Gomm.

x. viminalis=*S. x smithiana* Willd. (*S. geminata* Forbes). St Mary's: near Tremelethen, Grose, 1939; Old Town Marsh, Grose, 1952 det. R. D. Meikle; fairly common in Old Town Marsh and at Tremelethen, Howitt. Tresco: Duckery Dell, 1968 (spec. det. R. C. L. Howitt) R. Lancaster.

There has been considerable planting of willows in Scilly. Townsend, 1864, records that they were planted in the withy-bed at Tresco, while Woodley, 1822 advocated planting them by the Great Pool to make baskets for lobsters and crabs and implies that there were no willows there in his time. Porthellick, on St Mary's, is derived from Celtic roots meaning 'Cove of the Willows', but several authors suggested that willows ought to be planted there, which implied that there were few or none there at the time.

ERICACEAE

Rhododendron L.

R. ponticum L.   Common Rhododendron.   Naturalised alien. Heaths and cliff slopes.   Locally abundant.

M.-.T.-.-. : -.-.-.-.   First record: Lousley, 1967.

St Mary's: Bar Point, a few plants, 1963. Tresco: Gimble Porth, rapidly spreading from seed, 1939; Middle Down, spreading, 1952; Monument Hill, planted and spreading, 1952.

The introduction of this aggressive species into Scilly is to be regretted.

Calluna Salisb.

C. vulgaris (L.) Hull   Ling.   Native.

Heaths and downs, on shallow soils over granite, and on consolidated dune.   Abundant.

M.A.T.B.MN. : S.-.H.TN.E.   First record: Millett, 1852.

Ling is the dominant plant over much of the high ground, where the exposure to strong winds results in dwarfed stunted plants. Formerly it provided the 'peat' which was the main fuel of the islands and such names as 'Turfy Hill' commemorate allotments where it was dug.

Erica L.

E. tetralix L.   Cross-leaved Heath.   Error.

Recorded, on the authority of Augustus Smith, as 'common' by Townsend, 1864, and Ralfs copied this into his *Flora*. I have no evidence that it has occurred in the islands.

E. cinerea L.   Fine-leaved Heath.   Native.
Well drained places over granite on the heaths.   Common.
M.A.T.B.MN. : S.-.H.TN.E.   First record: Millett, 1852.
E. erigena R. Ross (*E. hibernica* (Hook. and Arn.) Syme, *E. mediterranea*
auct.)   Irish Heath.   Planted alien.
Tresco: established, May 1929. Bush 5 ft high. R. Meinertzhagen
(Herb. Mus. Brit.).

E. vagans L.   Cornish Heath.
Tresco: by the football field on Appletree Banks, 1966, P. Z.
MacKenzie. This is a cultivated form, and was no doubt planted
there from the Abbey Gardens.

PLUMBAGINACEAE
Armeria Willd.
 A. maritima (Mill.) Willd.   Thrift.   Native.
All round the coasts on rocks, shingle and sand, and ascending
to the higher ground where exposed to occasional salt-spray.
Abundant.
M.A.T.B.MN. : S.A.H.TN.E.   First record: Millett, 1952.
The abundance of Thrift is one of the botanical features of Scilly
and in May, when in full flower, it makes a colourful contribution
to the beauty of the headlands and smaller islands. White flowers
are sometimes found, as on Shipman Head, Bryher. Townsend,
1864 recorded *A. pubescens* Link from 'Annette and elsewhere' but
the hairiness of the calyx is no longer regarded as of importance.
Recently Miss C. W. Muirhead has drawn attention to var. *pubigera*
(Desf.) Boiss. with pubescent involucral sheaths and bracts which
occurs in the west of Ireland and Lusitania (*Proc. B.S.B.I.*, **6**, 279,
1966). She has examined my specimens of Thrift and points out
that those from Shipman Head, Bryher (444) have some hairs
on the involucre and bracts which suggests the influence of the
more densely pubescent var. *pubigera*.

PRIMULACEAE
Primula L.
 P. vulgaris Huds.   Primrose.   Native?
In a churchyard, under bracken on hillslopes and on consolidated
dune.   Very local.
M.-.T.-.-. : S.-.-.-.-.   First record: Townsend, 1864.

St Mary's: Ralfs *Flora;* Old Town Churchyard. Tresco: in plenty near Carn Near. Samson: in plenty above the lower ruins on South Hill, and one patch on lower slope, North Hill.

In spite of its abundance in Cornwall the status of the Primrose in Scilly is open to question. On Samson, an island now uninhabited, it grows in a natural habitat but is associated with the ruins. Scillonians visit Samson to gather the flowers and the roots in Old Town Churchyard probably came from here. On the consolidated dunes at Carn Near, garden plants have long been planted. Elsewhere in Scilly Primrose appears to be absent from the clays where it might be expected to grow as a native

Lysimachia L.
L. nummularia L.   Creeping Jenny.   Native.
Damp track in carr.   Very rare.
. -.T.-.-. : -.-.-.-.-.   First record: Lousley, 1967.
Tresco: Track from Abbey to Great Pool, 1953, B. T. Ward; seen again, no flowers, 1963.

L. vulgaris L.   Yellow Loosestrife.   Established alien.
Wet elm copse on moors.   Very rare.
M.-.-.-.-. : -.-.-.-.-.   First record: Lousley, 1967.
St Mary's: copse on Lower Moors, north-west end, 1967, P. Z. MacKenzie (862). This is near the ruins of a hut, and was probably planted.

Anagallis L.
A. tenella (L.) L.   Bog Pimpernel.   Native.
Marshes, tracks, poolsides and dry heath.   Occasional.
M.A.T.B.-. : -.-.-.-.-.   First record: Millett, 1852.
St Mary's: 1872 and 1877, Herb. Beeby; Holy Vale, Grose; Old Town Marsh, Herb. Dallas; Lower Moors near Rose Hill; track near shore, Helvear Down. St Agnes: near Tinflat Point, Townsend, 1864. Tresco: about Abbey Pool; in dry Callunetum, Merchant's Point. Bryher: North end, M. Walpole, 1969.

In the humid atmosphere of Scilly this grows in much drier places than are usual on the mainland.

A. arvensis L. subsp. *arvensis*   Scarlet Pimpernel.   Native.
Cultivated ground, roadsides, cliffs and consolidated dune.
Common.
M.A.T.B.MN. : S.-.H.TN.E.   First record: Millett, 1852.
var. *carnea* Schrank.   Plants with flesh-coloured flowers and
suberect stems occur in widespread colonies and are not restricted to
dunes. St Mary's: Garrison Hill, 1924, Miss E. Farthing comm. L. J.
Tremayne; roadside on Garrison. St Agnes: Troy Town and
Wingletang Downs (433). Tresco: plentiful on sand-dunes, Carn
Near, Lousley in *B.E.C. Rep.*, **11**, 267, 1937. Bryher: Rushy Bay.
St Martin's: coast in various places such as Lower Town and
Porth Seal. Rilstone found that in Cornwall plants with flesh-
coloured flowers were almost exclusively maritime (*J. Bot.*, **76**, 85,
1938) and D. E. Allen pointed out that the reason for the prevalence
of the flesh-coloured form in the milder, westerly coastal areas is
obscure (*Proc. B.S.B.I.*, **1**, 156–157, 1954). All the places where it
grows in Scilly are likely to have high concentrations of sea-salt. Here
also may belong specimens from Peninnis Head, St Mary's, collected
by Grose and determined as var. *pallida* by E. M. Marsden-Jones.
Plants with blue flowers (*A. arvensis* L. subsp. *arvensis* forma
azurea Hyland.) occur occasionally. A specimen from garden
ground, St Mary's, July 1890 (Herb. Edinb.) was recorded as
A. caerulea Schreb. in Somerville, 1893, and similar blue-flowered
forms were found in abundance at Newford, St Mary's by R. C. L.
Howitt in 1962, and a single plant was found in a bulbfield on
Tresco by Miss M. B. Gerrans in 1963 (Herb. Mus. Brit.).

A. minima (L.) E. H. L. Krause (*Centunculus minimus* L).   Chaff-
weed.   Native.
Damp and usually sandy places.   Rare.
M.A.T.-.-. : -.-.-.-.   First record: Townsend, 1864.
St Mary's: Old Town Marsh, 1862, Herb. Townsend. St Agnes:
Priglis Bay, 1948, Wanstall; damp track in centre of island, 1952,
Grose. Tresco: by Abbey Pool.

Glaux L.
G. maritima L.   Sea Milkwort.   Native.
By brackish pools, marsh ditches, and under coastal rocks.
Rather rare.
M.A.T.B.-. : -.-.-.-.   First record: Millett, 1852.

St Mary's: Hugh Town and Old Town Marsh, Townsend; Higher Marsh, Ralfs' *Flora*; by Porthellick Pool (146); under coastal rocks below Giant's Castle. St Agnes: in several places, Grose; Gugh, Dallas; Guinea Money Carn, Ribbons and Wanstall. Tresco: plentiful around Abbey Pool; Duckery. Bryher: plentiful round the pools.

Samolus L.
S. valerandi L.   Brookweed.   Native.
Marshes, pools and damp places on cliffs.   Rare.
M.A.T.-.-. : -.-.-.-.   First record: Millett, 1852.
St Mary's: Lower and Higher Moors. St Agnes: Gugh and Horse Point, Dallas; Dropmore Point, Gugh, Ribbons and Wanstall; below Carn Wrean, Gugh. Tresco: Lower Pool, Grose; by Abbey Pool.

OLEACEAE

Fraxinus L.
F. excelsior L.   Ash.
Has been determined from charcoal samples in second–fourth century remains at Halangy (Ashbee, 1955, 197), and was planted about 1650 at Holy Vale (Turner in *The Scillonian*, **159,** 154, 1964), but not known to grow in Scilly in recent years.

Ligustrum L.
L. vulgare L.   Wild Privet.   Native.
Consolidated dune, but also on slopes over granite.   Locally common.
M.-.T.B.MN. : S.-.-.-.   First record: Millett, 1852.
On all the larger islands except St Agnes. This species is subprostrate and especially plentiful on the dunes at Bar Point, St Mary's, Gimble Porth and Appletree Banks, Tresco, and about Great and Little Bays on St Martin's.

L. ovalifolium Hassk.   Japanese Privet.   Planted alien.
In hedges and on the coast, sometimes not obviously planted. Local.   M.-.-.-. : -.-.-.-.   First record: Lousley, 1967.
St Mary's: Holy Vale; Maypole; Pelistry and Toll's Island (755); Watermill Cove, etc. Often confused with *L. vulgare*.

APOCYNACEAE
Vinca L.

V. major L.   Greater Periwinkle.   Denizen.

Hedgerows, by tracks, dunes, and old quarries.   Rare.

M.A.T.B.-. : -.-.-.-.-.   First record: Smith, 1909.

St Mary's: an escape, Smith, 1909; pit near Woolpack Point, 1953; now spread along track to coastguards' houses, 1963; Rosehill, 1956; near Airport, 1953, D. McClintock; pit north of Porthloo, 1959. St Agnes: near Parsonage, 1956. Tresco: established on dune near New Grimsby, 1953; Appletree Banks 1968, R. Lancaster. Bryher: in two places near cottages, 1953. This species has spread considerably during the last 15 years.

GENTIANACEAE
Centaurium Hill

C. latifolium (Sm.) Druce This species, known only from the Lancashire dunes where it is now extinct, was recorded in error by Townsend, 1864. (see below).

C. erythraea Rafn   Common Centaury.   Native.

Dunes and other sandy places, walltops, downs, and sea-cliffs. Common.

M.A.T.B.MN. : S.-.H.TN.E.   First record: Millett, 1852.

Varies widely according to available moisture and exposure to wind. An extreme almost stemless form (*subcapitatum*) is fairly frequent in exposed places, and fasciculate and other forms have also been given varietal names. White flowers are frequent (C. Nicholl in *The Scillonian*, **27,** 274, 1952.)

C. capitatum (Willd.) Borbás This 'species', which differs only from the subcapitate form of *C. erythraea* in the insertion of the stamens at the base of the corolla-tube, was collected on sands at Tresco by R. B. Ullman in 1911 (Herb. Lousley). It occurs also in Cornwall and Guernsey but its taxonomic value is doubtful.

C. littorale (D. Turner) Gilmour was recorded by Millett, 1852 and by Townsend, 1864. Specimens in Townsend's herbarium are *C. erythraea* and he recognised his own error by writing in a copy of his paper 'I am inclined to refer these (*E. latifolia* Sm. and *E. littoralis* Fries) to *E. centaurium* judging from specimens in my herbarium'. Although the species is given for St Mary's in Ralfs' *Flora*, it is unlikely that it has ever occurred as far south as Scilly.

BORAGINACEAE
Symphytum L.

S. officinale L. was recorded from Hugh Town, St Mary's and St Martin's by Smith, 1909, and there is a specimen so determined by Edinburgh in Herb. Dorrien-Smith. There is little doubt that all three records belong to the next entry.

S. x uplandicum Nyman   Russian Comfrey.   Established alien.
Damp roadsides and cultivated ground.   Local.
M.-.T.-.-. : -.-.-.-.-.   First record: Lousley, 1939.
St Mary's: Watermill Cove; near Tremelethen (148); Normandy; Green, Grose; Carn Friars. Tresco: near Great Pool, 1939, Grose; Abbey Grounds; abundant along Pool Road; kitchen garden and common along Pool Road, Herb. D.-S.
Russian Comfrey has long been planted as a forage plant and once introduced is extremely difficult to eradicate.

Borago L.
B. officinalis L.   Borage.   Established alien.
Waste ground and places where rubbish is thrown out from gardens.   Rare.
M.A.T.-.MN. : -.-.-.-.-.   First record: Townsend, 1864.
St Mary's: waste ground east of Hugh Town, Townsend; bulbfield between Porthloo and Rocky Hill, U. Duncan. St Agnes: Ralfs' *Flora*; Boyden in Tonkin and Row, 1893; in the village Howitt. Tresco: rubbish-heap, Appletree Banks. St Martin's: near the Church, M. Knox; Higher Town, Dallas.

Pentaglottis Tausch
P. sempervirens (L.) Tausch   Evergreen Alkanet.   Established alien.
Sand-dune where garden rubbish may have been dumped.   Very rare.   -.-.T.-.-. : -.-.-.-.-.   First record: Lousley, 1939.
Tresco: Appletree Banks (154), and seen there at intervals since.

Anchusa L.
A. azurea Mill.   Garden outcast.
A single plant was seen by the sea near Old Town, St Mary's by Miss M. McCallum Webster in June 1966 and was also reported by P. Z. MacKenzie. This was at Porth Minick, formerly used for refuse tipping, and I saw two plants there in May 1967.

Lycopsis L.
L. arvensis L.   Small Bugloss.   Native.
Cultivated ground and sandy places near the shore.   Very
common.
M.A.T.B.MN. : S.-.-.-.-.   First record: Millett, 1852.
On Samson, where there has been no cultivation for over a century, it
occurs as a very small state on the sandy bar which joins the two hills.

Myosotis L.
M. scorpioides L.   Water Forget-me-not.   Native.
Very rare.
-.-.T.-.-. : -.-.-.-.-.   First record: Boyden in Tonkin and Row, 1893.
Tresco: one place by Great Pool, 1939, Grose; there is a specimen
in Herb. D.-S. named at Edinburgh.

M. secunda A. Murr.   Creeping Forget-me-not.   Native.
Marshes, ditches and pool sides.   Frequent.
M.-.T.-.-. : -.-.-.-.-.   First record: Townsend, 1864.
St Mary's: Townsend, 1864; 1873, Beeby in Herb. S.L.B.I.;
Higher Marsh, Ralfs' *Flora*; Higher Moors, Smith 1909; Holy Vale
and Old Town Marsh (both det. A. E. Wade); roadside gutter,
Tremelethen (156); Higher Moors (677); Watermills (158); by
Porthellick Pool in Phragmites (750); roadside pond, Lower
Newford. Tresco: Townsend, 1864; by Abbey Pool (699); Great
Pool.
M. caespitosa K. F. Schultz.   Recorded by Boyden in Tonkin and
Row, 1893, by Somerville, 1893 and by me in *J. Bot.*, 1939, 155. My own
specimens were *M. secunda* and the earlier records were probably due to
similar confusion.

M. arvensis (L.) Hill   Field Scorpion-grass.   Native.
Cultivated fields, hedges, etc.   Common.
M.A.T.B.MN. : S.-.-.-.-.   First record: Townsend, 1864.
On Samson it grows on the sandy isthmus. Plants with white
flowers occur in a field near Hugh Town.

M. discolor Pers.   Changing Scorpion-grass.   Native.
Cultivated ground, cliffs, heaths, etc.   Rather common.
M.A.T.B.MN. : S.-.-.-.E.   First record: Townsend, 1864.
Found on Samson and Great Ganilly by Miss M. Jaques in 1957.
Plants with white flowers (and usually paler foliage) have been

found by Ralfs on the 'Battery Ground, St Mary's', and by me on Samson Hill, Bryher and Appletree Banks, Tresco.

M. ramosissima Rochel (*M. collina* auct.)   Early Scorpion-grass. Native.

Dunes, sandy places on heaths, paths and cliffs.   Common.

M.A.T.B.MN. : S.AT.H.TN.-.   First record: Townsend, 1864.

Echium L.

E. vulgare L.   Viper's Bugloss.   Casual.

Sandy waste ground near shore.   Very rare.

-.A.-.-.-. : -.-.-.-.-.   First record: Lousley, 1939.

St Agnes: near Gugh Farm on a rubbish-tip, 1933, Herb. D.-S.; in small quantity at same place, 1939; several plants in sandy field above Gugh Bar, 1969, R. Lancaster. Like *Reseda lutea* and *Melilotus officinalis* which appeared at the same place about 1933, this must be regarded as an introduction arising from some agricultural activity.

E. lycopsis L. (*E. plantagineum* L.)   Jersey Bugloss.   Native.

Rough ground.   Very rare.

-.-.-.-.MN. : -.-.-.-.-.   First evidence: Scilly Isles, L. Mann, 1882, Herb. Univ. Cantab.   First record: Lousley, 1967.

St Martin's: 1956, Howitt; 1957, 12 plants; and later years. In Scilly this species grows under conditions very similar to those where I have seen it in Jersey—or indeed in the Mediterranean. It is biennial, which may explain why I and others failed to find it earlier.

Tree Echiums. These handsome plants from the Canaries grow freely at Tresco Abbey and in many gardens throughout the islands. Those at Tresco are said to be *E. pininana* Webb and Berth., *E. auberianum* Hort., *E. fastuosum* Jacq., *E. callithrysum* Webb ex C. Bolle, and *E. x 'scilloniensis'*, but I am unable to trace that the

---

Fig 10 Succulents: 1 *Oscularia deltoides* (leaf and leaf section x1). 2 *Erepsia heteropetala* (leaf x1). 3 *Disphyma crassifolium* (leaf x1½). 4 *Carpobrotus acinaciformis* (leaf section x½). 5 *Drosanthemum floribundum* (leaf x1½)

2

5

last-mentioned has been validly published. I suspect that some of these spread from self-sown seed at, for example, Appletree Banks and Holy Vale, but cannot claim that they are established. Miss W. Frost reports that in 1967 she found that '*scilloniensis*' had seeded and spread in many places, and especially in the moat of the Star Castle Hotel.

CONVOLVULACEAE

Convolvulus L.
  C. arvensis L.   Field Bindweed.   Native.
Cultivated and waste ground, roadsides.   Common.
M.A.T.B.MN. : -.-.-.TN.-.   First record: Millett, 1852.
C. althaeoides L. 'Sent as wild from Scilly', July 1883, A. Bennett, ex-herb. W. F. Miller, Herb. Bristol Museum. I am grateful to D. McClintock for drawing my attention to this specimen. There are no later records and the plant was probably a casual grain alien, or perhaps thrown out from Tresco Gardens, or Sicily may have been intended.

Calystegia R.Br.
  C. sepium (L.) R.Br.   Hedge Bindweed.   Native.
Hedges, thickets, and scrambling over taller vegetation in waste places.   Frequent.
M.A.T.B.MN. : -.-.-.TN.-.   First record: Millett, 1852.
Subsp. *roseata* Brummitt); Porth Minick; Old Town Hill as *C. pulchra*, 1966, P. Z. MacKenzie. This is plentiful about Lower Moors and is no doubt the plant with handsome pink flowers which Townsend found in 'Heugh Town Marshes' too inaccessible to collect (Townsend, 1864). The sub-species has a west coast distribution and is found on the opposite coast of Cornwall in Marazion Marsh.

  C. silvatica (Kit.) Griseb.   American Bellbine.   Established alien.
Hedges and disused quarry.   Very local.
M.-.-.-.-. : -.-.-.-.-.   First record: Lousley, 1967.
St Mary's: Near houses, Holy Vale and Porthloo, 1952; quarry at junction of Town Lane and Telegraph Road, 1956.

  C. soldanella (L.) R.Br.   Sea Bindweed.   Native.
Sandy shores and dunes.   Very common.
M.A.T.B.MN. : S.-.H.TN.E.   First record: Millett, 1852.

Cuscuta L.

C. epithymum (L.) L.   Heath Dodder.   Native.

Parasitic on gorse.   Very rare.

-.-.T.-.-. : -.-.-.-.-.   First record: Townsend, 1864.

Tresco: On gorse near and above the Abbey, Townsend; Monument Hill, 1952.

SOLANACEAE

Nicandra Adans.

N. physalodes (L.) Gaertn.   Shoo-fly Plant.   Naturalised alien.

Weed in bulbfields.   Locally common.

M.-.T.-.-. : -.-.-.-.-.   First record: Miss Anne Dorrien-Smith in Thurston, 1936.

St Mary's: near Lower Moors, 1968, P. Z. MacKenzie. Tresco: Plentiful in bulbfields, Anne Dorrien-Smith, Herb. Thurston; Weed of cultivation, Borough, 1953, Polunin; weed in Abbey grounds (349), 1939, 1956: abundant by Pool Road in crops (518), 1952, 1956, 1970.

Hyoscyamus L.

H. niger L.   Henbane.   Native.

Sandy shores and dunes, cultivated ground.   Rare.

M.A.T.-.-. : -.-.-.TN.-.   First record: Millett, 1852.

St Mary's: near Hugh Town, Townsend; Porthcressa, 1963; Airport, 1939, Grose. St Agnes: The Gugh, 1952, Grose. Tresco: Farm Bank, Herb. D.-S.; abundant by Abbey Farm, 1956 and again, 5 ft tall, 1957; beach near Carn Near, G. Andrews; New Grimsby, 1939. Tean: 1939, Grose.

Solanum L.

S. dulcamara L.   Bittersweet.   Native.

Roadsides, cliffs and shores.   Very common.

M.A.T.B.MN. : S.AT.H.TN.E.   First record: Millett, 1852.

S. nigrum L.   Black Nightshade.   Native.

Roadsides, rubbish heaps, and shores.   Frequent.

M.A.T.B.MN. : S.AT.-.-.E.   First record: Lawson, 1870.

As in Cornwall, this is much rarer than the last species. 'Susan Bates, aged 2, daughter of a gardener on Tresco, died on July 18, 1966 due to eating 'deadly nightshade' berries revealed by post-mortem on St Mary's', *Scillonian*, **167**, 105, 1966.

S. sarrachoides Sendtn.   Green Nightshade.   Naturalised alien.
Sandy bulbfields.   Locally plentiful.
-.-.-.-.MN. : -.-.-.-.-.   First record: Lousley, 1967.
St Martin's: In great abundance below Middle Town and towards Lower Town as weed in bulbfields, 1952 (523). A few plants have corollas with a persistent purple flush (524).

S. laciniatum Ait.   Kangaroo Apple.   Naturalised alien.
Sand-dune, waste ground, roadsides.   Rare.
-.-.T.-.-. : -.-.-.-.-.   First record: Hosking in Thurston and Vigurs, 1922.
Tresco: Sandy shore, September 1920 A. J. Hosking in Herb. Thurston; near Abbey Farm, 1956 onwards. This was first recorded as *S. aviculare* Forst., a New Zealand species, which has been confused with the mainly Australian *S. laciniatum*. The freedom with which this grows from self-sown seed in Abbey Gardens was reported by E. Brown, *Gdnrs'. Chron.*, **98,** (ser. 3), 102–103, 1935.

S. tuberosum L. Potato plants are common on the shores of all the inhabited islands, in places where rubbish is dumped. Owing to the mild winters they may sometimes persist from year to year for a time.

Datura L.
D. stramonium L.   Thorn Apple.   Established alien.
Rubbish heaps and cultivated ground.   Rare.
-.A.T.-.-. : -.-.-.-.-.   First record: Lousley, 1967.
St Agnes: field on Capt. Dick Legg's Farm, 1948, P. J. Wanstall. Tresco: Abbey Gardens, 1938, G. Andrews, and again 1956; waste ground by Abbey Farm, 1956.

SCROPHULARIACEAE

Verbascum L.
V. thapsus L.   Great Mullein.   Native.
Roadsides, walls, dunes and quarries.   Not common.
M.A.T.B.MN. : -.-.H.TN.-.   First record: Millett, 1852.
Too widespread to justify giving localities, but in each there are very few plants and, being biennial, these are sometimes represented by barren rosettes of leaves.

V. phlomoides L.   Casual, known only from St Mary's where
P. Z. MacKenzie found the white-flowered form as a weed in a
garden opposite the Vicarage on Tremellyn Hill, and the more
usual yellow-flowered plant beside the new dump on Lower Moors.

V. lychnitis L. Casual. Recorded from Tresco by Miss Anne
Dorrien-Smith in Thurston and Vigurs, 1929.

V. nigrum L.   Dark Mullein.   Native?
Sand-dunes.   Very rare.
-.-.T.-.-. : -.-.-.-.-.   First record: Smith, 1909.
Tresco: sandy flats, Smith loc. cit.; seen again on Tresco by
H. Downes about 1921.

V. blattaria L.   Moth Mullein.   Casual.
Recorded by Miss Anne Dorrien-Smith in Thurston and Vigurs,
1928 from 'Tresco. Most of the plants in the garden, on the tops
of walls, and under trees'. This suggests that it was established but
there are no subsequent records.

V. virgatum Stokes.   Twiggy Mullein.   Casual.
Tresco: Appletree Banks, 1956.

Misopates Raf.
M. orontium (L.) Raf.   Lesser Snapdragon.   Colonist.
Cultivated fields and gardens.   Uncommon.
M.A.T.B.MN. : -.-.-.-.-.   First record: Bree in Forbes, 1821.
St Mary's: Ralfs' *Flora*: Smith, 1909. St Agnes: Ralfs' *Flora;*
Smith, 1909. Tresco: Smith, 1909; plentiful and very fine in Abbey
Gardens (355). Bryher: Bree, op. cit.; Southward. St Martin's:
Ralfs' *Flora*; Smith, 1909; Downes, 1921; above Higher Town Bay
and below Lower Town.

Linaria Mill.
L. vulgaris Mill.   Yellow Toadflax.   Native.
Very rare.
-.-.T.-.-. : -.-.-.-.-.   First record: Lawson, 1870.
Tresco: Lawson. Also recorded from Scilly by Ralfs in Davey,
1909.

Kickxia Dumort.

K. spuria (L.) Dumort.   Round-leaved Fluellen.
Recorded from Tresco, as rare, by Townsend, 1864.

K. elatine (L.) Dumort.   Sharp-leaved Fluellen.   Native?
Very rare.
M.A.T.-.-. : -.-.-.-.-.   First record: Townsend, 1864.
St Mary's: Ralfs' *Flora*. St Agnes: 1966, Miss H. M. Quick.
Tresco: rare?, Townsend; Ralfs' *Flora;* Pool Road, 1970. There
is an unlocalised specimen in Herb. Dorrien-Smith collected by
Mrs Perrycoste.
Fluellens are characteristically weeds in cereal crops on light soils.
They probably decreased when the less suitable conditions of
bulbfields replaced cereals.

Cymbalaria Hill
C. muralis Gaertn., Mey. and Scherb.   Ivy-leaved Toadflax.
Denizen.
Old walls and stone hedges.   Rare.
M.-.T.-.MN. : -.-.-.-.-.   First record: Lawson, 1870.
St Mary's: walls of Star Castle; Holy Vale (*flore albo*). Tresco:
Lawson; near Abbey, Grose; Abbey Grounds; New Grimsby.
St Martin's: 1938, M. Knox; Higher Town.

Scrophularia L.
S. nodosa L.   Common Figwort.   Native.
Hedgebanks.   Rare.
M.-.T.-.-. : -.-.-.-.-.   First record: Townsend, 1864.
St Mary's: Ralfs' *Flora*; Lane near Higher Trenoweth, scarce (285).
Tresco: Lane near Borough, scarce.
S. auriculata L. (*S. aquatica* auct.). Water Figwort. Error? This was
recorded by Townsend, 1864 as 'Not so common as *S. nodosa*'. It appears
as *S. balbisii* from St Mary's in Ralfs' *Flora*. Boyden included it in his
list in Tonkin and Row, 1893, but as he gives no other species of *Scrophu-*
laria, this is clearly an error. Many recent visitors have also claimed it,
but in every case it has been established that the plant they found was
S. scorodonia, a species which varies considerably and with which most
visiting botanists are not familiar. *S. auriculata* is a not unlikely species to
occur, but until fresh evidence can be produced, all the records must be
regarded as errors.

S. scorodonia L.   Balm-leaved Figwort.   Native.

Hedgerows, marshes, sand-dunes in *Ammophiletum*, and rocky shores.   Very common.

M.A.T.B.MN. : S.-.H.-.E.   First record: Bree in Forbes 1821, from 'Within the ruins of the Abbey, Tresco Island'.

This species is abundant on St Mary's and Tresco, and common on the other three inhabited islands. It is plentiful on Samson and St Helen's and in less quantity on Great Ganilly. Few species thrive in such a wide range of habitats.

Mimulus L.

M. guttatus DC.   Monkey-flower.

Found by P. Z. MacKenzie in 1967 in a copse in Lower Moors, St Mary's (863). This alien is likely to become established.

Limosella L.

L. aquatica L.   Mudwort.   Native.

Muddy shore of a pool frequented by ornamental fowl.   Very rare.   -.-.T.-.-. : -.-.-.-.-.   First record: Gibbs, 1884, see below. Tresco: Lake in front of Abbey, A. E. Gibbs in *Rep. Bot. Rec. Cl.*, **1883**, 18, 1884; Downes, 1921. F. J. Hanbury claimed to have found it in the same place in 1923, but his specimen is *Elatine hexandra*, Herb. Mus. Brit.

Sibthorpia L.

S. europaea L.   Cornish Moneywort.   Native.

Sides of ditches and moist shady banks.   Rare.

M.-.-.-.-. : -.-.-.-.-.   First record: Bree, 1831.

St Mary's: Abundant on the banks of a rivulet on the road leading from Holy Vale towards Hugh Town. Banks of a ditch in the orchards near Old Town Marsh, Townsend, 1864; Lower Moors near Low Pool, Dallas; shore at Tregear's Porth (261); corner of field at Watermill Cove; abundant in ditches bordering Watermill and Pungies' Lanes (289); Porthellick, 1955, Howitt.

Digitalis L.

D. purpurea L.   Foxglove.   Native.

Roadsides, hedgebanks, cliff slopes, etc.   Very common.

M.A.T.B.MN. : S.-.H.TN.E.   First record: Millett, 1852.

Veronica L.

V. beccabunga L.   Brooklime.   Native.
Recorded by Boyden in Tonkin and Row, 1893, and by Tellam in Davey, 1909. There is no reason to doubt the records, but the locality is not known.

V. officinalis L.   Common Speedwell.   Native.
Heaths, and under bracken on the downs and cliffs.   Frequent.
M.A.T.B.MN. : S.-.-.-.-.   First record: Townsend, 1864.

V. montana L.   Wood Speedwell   Native.
Shady place.   Very rare.
-.-.T.-.-. : -.-.-.-.-.   First record: Somerville, 1893.
Tresco: Weed in Abbey Gardens, 1968 (spec.), R. Lancaster. This locality is on, or near, the site of Abbey Wood and, with other woodland species still there, it is likely to be a relic of former conditions.

V. chamaedrys L.   Germander Speedwell.   Native.
Hedgebanks, roadsides, under bracken on the cliffs, etc. Common.
M.A.T.B.MN. : S.-.H.TN.E.   First record: Millett, 1852.

V. serpyllifolia L.   Thyme-leaved Speedwell.
Native.   Damp pastures and roadsides.   Rare.
M.A.T.B.MN. : -.-.-.-.-.   First record: Townsend, 1864.
St Mary's: Ralfs in litt to Townsend, 1876; 1939, M. Knox. St Agnes: 1960. E. G. Philp. Tresco: Herb. D.-S. Bryher: damp pasture near Pool, 1952, A. Conolly; 1957, M. Jaques. St Martin's: Higher Town, 1953, Polunin.
Townsend gave this species as 'common', but I have not yet seen it in Scilly.

V. arvensis L.   Wall Speedwell.   Native.
Roadsides, dunes, walls, and cultivated ground.   Common.
M.-.T.B.MN. : S.-.-.-.E.   First record: Townsend, 1864.
On dunes this is often 2–3 cm tall (var. *nana* Poir.), in bulbfields sometimes luxuriant.

V. hederifolia L.   Ivy-leaved Speedwell.   Native.

(*top*) Sea Storksbill, *Erodium maritimum*, abundant on bare ground; (*bottom*) Sea Cottonweed, *Otanthus maritimus*, now extinct, photographed in September, 1966 on St Martin's, shortly before it disappeared.

Cultivated land and waste places.   Frequent.
M.A.T.B.MN. : -.-.-.-.-.   First record: Ralfs in Marquand, 1893.
The late date of the first record is explained by the early flowering so that it would be unlikely to be found by Millett, Townsend, etc.

V. persica Poir.   Common Field Speedwell.   Colonist.
Cultivated ground, roadsides and waste places.   Abundant.
M.A.T.B.MN. : -.-.-.-.-.   First record: Perrycoste in Davey, 1909.
Perrycoste observed this species on St Mary's in 1898 and Smith on the same island in 1906. It seems that this alien first reached Scilly towards the end of the nineteenth century, spreading rapidly, first on St Mary's and then to the other inhabited islands. On St Martin's my notes indicate a great increase between 1939 and 1953.

V. polita Fr.   Grey Field Speedwell.   Native.
Cultivated ground.   Rare.
-.A.T.-.-. : -.-.-.-.-.   First record: Lousley, 1967.
St Agnes: near Middle Town, 1952, Grose. Tresco: weed in Abbey Gardens, 1956 (722).

V. agrestis L.   Green Field Speedwell.   Native.
Gardens and rich cultivated ground.   Very rare.
M.A.T.-.-. : -.-.-.-.-.   First record: Townsend, 1864.
St Mary's: Ralfs' *Flora*. St Agnes: Ralfs' *Flora*; 1966, H. M. Quick. Tresco: Abbey Gardens, 1960, J. Russell.

V. filiformis Sm.   Established alien.
St Mary's, 1955, Howitt.

Hebe Commers.
H. lewisii (Armstrong) Cockayne and Allen   Hedge Veronica.
Naturalised alien.
Planted as a windbreak and self-sown on walls and cliffs. (Fig 9, p 185.)
M.A.T.B.MN. : -.-.-.-.-.   First record: Lousley, 1939.
This New Zealand shrub with its tough leathery leaves and purple flowers is familiar to visitors as the most commonly planted windbreak. It grows freely from seed but the seedlings usually have

paler leaves than the parents which are garden hybrids. Self-sown plants are common—easily found examples being on the gateway to the parish church, Hugh Town, and a wall at Porthloo, St Mary's, and on the cliff near Brandy Point, St Martin's. Donkeys prefer Veronica to oats, and it is cut for them to eat (*Scillonian*, **20**, 174, 1946).

Pedicularis L.

P. palustris L. This was recorded by the Misses Millett, 1852 but as they fail to mention *P. sylvatica*, which is very common, they were probably in error. Ralfs gives both species, and cites St Mary's and Tresco for *P. palustris* in his *Flora*. It is also given in Somerville, 1893. Higher Moors was a possible habitat but these two authors made so many errors that I hesitate to accept the species for Scilly.

P. sylvatica L.   Lousewort.   Native.
Heathy ground, mainly on the downs.   Very common.
M.A.T.B.MN. : -.-.-.-.   First record: Townsend, 1864.
Townsend pointed out that in Scilly this species grows on 'the most barren and exposed heaths', the high ground of the downs where the soil is much drier than in the usual mainland habitats. It also grows on Appletree Banks on consolidated dune. Miss B. M. C. Morgan found a plant with white flowers on the Golf Links, St Mary's.

Euphrasia L.

E. officinalis L. *sensu lato*   Eyebright.
The aggregate species was first recorded by Millett, 1852 and Eyebrights are very common on all the inhabited and the larger uninhabited islands. The following account is based on determinations by the late H. W. Pugsley with the additions, as stated, by P. F. Yeo.

E. micrantha Reichb. St Mary's: Porthellick, 1939, Grose—the only record.

E. curta (Fr.) Wettst. St Mary's: Bar Point—f. *glabrescens* (282), Lousley, 1940.

E. tetraquetra (Bréb.) Arrondeau (*E. occidentalis* Wettst.).   Native.
Grassy cliff slopes, downs, sand-dunes, tops of stone hedges. Abundant.

228　　SCROPHULARIACEAE

M.A.T.B.MN. : -.-.-.-.-.　　First record: Townsend, 1897.

This early flowering Eyebright is by far the most common representative of the genus in Scilly. It occurs also on some of the uninhabited islands but material has not been checked. Specimens collected on St Mary's by Townsend in 1862 were named by Wettstein and were included in those cited in the first record of the species for Britain (Townsend, *J. Bot.*, **35,** 419, 1897). It is a characteristic species of south-west England, occurring abroad only in north-west France.

Pugsley also determined var. *praecox* Bucknall from Bryher: near Samson Hill (160 and 161)—var. *minor* Pugsley from Bryher: Samson Hill (331) and var. *calvescens* Pugsley from St Mary's: Salakee Downs (162); Golf Links (163) and amongst Ling, Peninnis Head (423). The first two appear to be only habitat forms.

E. nemorosa (Pers.) Wallr. To this segregate Davey, 1909, placed Millett's record of *E. officinalis* without justification. The species is unknown in Scilly.

E. confusa Pugsl. St Mary's: By track, Peninnis Head—f. *albida* 'unusually robust' (331), Lousley, 1940. To this Pugsley was also inclined to assign material from the Gugh (165).

E. brevipila Burnat and Gremli. St Mary's: C. E. Salmon in *J. Roy. Inst. Cornw.* 52, 1906 and Davey, 1909; Carn Morval Down, 1954 (671 and 673) det. P. F. Yeo. These are two habitat forms of the same purple-flowered plant—one drawn up under bracken, the other in the open, stouter with short internodes. Dr Yeo points out that Pugsley determined similar plants as *E. confusa.*

E. anglica Pugsley var. *gracilescens* Pugsley. St Mary's: Salakee Downs, 1939 (295), Lousley, 1940. This interesting slender plant grew under bracken by the path descending to Porthellick.

Odontites Ludw.

O. verna (Bellardi) Dumort. Recorded from St Mary's by Ralfs in litt. to Townsend, 1876, and repeated in Davey, 1909. In view of the many errors made by Ralfs, the record cannot be accepted without confirmation.

Parentucellia Viv.

P. viscosa (L.) Caruel　　Yellow Bartsia.　　Native.

Marshes, roadsides, damp sandy ground, bulbfields.　　Common.

M.A.T.B.-. : -.-.-.-.-.　　First record: Bree, 1831.

Common on St Mary's and Tresco, local on Bryher and St Agnes, and I have failed to find it on St Martin's. Sometimes in places which are apparently dry, as on roadsides and slopes on The Garrison.

OROBANCHACEAE

Orobanche L.

O. elatior Sutton. Recorded for 'Scilly' by Dr J. B. Montgomery in *Rep. Penzance Nat. Hist. and Antiq. Soc.*, **2**, 218, 1854, but in error. See Davey, 1909.

O. minor Sm.   Lesser Broomrape.   Native.
Parasitic on various species.   Rare.
M.A.T.-.-. : -.-.-.-.-.   First record: Millett, 1852.
St Mary's: Sandy field near the sea, E. of Hugh Town, Townsend, 1864; Porthloo, 1938; Porth Mellin, 1928, Herb. D.-S.; Porth Mellin, on carrot, 1938 (151); on imported *Senecio* in garden of new house on The Garrison, 1967, Alf Page. St Agnes: 1966, H. M. Quick. Tresco: Tresco Gardens, Herb. D.-S.

O. hederae Duby   Ivy Broomrape.   Native.
Parasitic on *Hedera helix*.   Very rare.
-.-.T.-.-. : -.-.-.-.-.   First record: Lousley, 1967.
Tresco: Tresco Gardens, 1937, 'but there many years', Herb. D.-S.; on ivy, Abbey Grounds, 1938 (100); Tresco gardens on various garden plants, 1957.

O. maritima Pugsl. A broomrape with the stem and corolla tinged with dull violet, which still grows on consolidated dune at Porth Mellin, has long puzzled collectors who have recorded it sometimes as *O. minor* and sometimes as '*O. amethystea*', which is *O. maritima* of current works. This violet-coloured plant grew formerly on sandy ground from Porth Cressa through Porth Mellin to Porthloo, but its habitats are now much restricted by building. It has been found parasitic on *Trifolium minus*, *T. repens*, *Eryngium maritimum*, *Daucus carota*, and doubtless other species.
It has appeared as '*O. amethystea*' in Ralfs' *Flora*, 1879; Somerville, 1893; Thurston and Vigurs, 1922; and *Rep. B.E.C.*, **7**, 399, 1924. There are specimens collected by Townsend, Downes and Curnow in Herb. S.L.B.I., by G. A. Holt in Douglas Museum; by Cunnack in Herb. Essex Field Club; and by myself, and as pointed out by Townsend (*MS.*, 1862) the bracts are too long and the corolla too arched for '*amethystea*'. The plant is a form of *O. minor*.

ACANTHACEAE
Acanthus L.

A. mollis L.   Bear's Breech.   Established alien.

Field-bank, roadsides, dunes, sandpit, and shore.   Local. (Plate p 289.)

M.A.T.-.-. : -.-.-.-.-.   First record: J. P. Mayne, 1851—see below.

St Mary's: roadside, Rocky Hill, 1939, Grose. Here it is persistent, competing with bracken; established on shore at Porthloo, 1956 onwards; 'two plants, probably escapes', Smith, 1909. St Agnes: J. P. Mayne ex Babington, *Ann. Mag. Nat. Hist.*, **8**, (ser. 2), 505, 1851 and *Phytologist*, **4**, 408, 1852; Tellam in Davey, 1909; refound by B. W. Ribbons, *Watsonia*, **2**, 392–393, 1953; still there. Tresco: 'wild on Tresco', 'Vagabondo', *Gdnrs' Chron.*, **26**, (N.S.), 558, 1886; established in the lower part of Abbey Gardens, on dunes and in a sandpit.

The St Agnes colony dates from *c* 1800 and, as Ribbons has described, attracted great interest, and found a place in national floras for over sixty years. The plants are not visible from roads or public paths and it is not surprising that so many botanists failed to refind it. *Acanthus mollis* is easily established and difficult to eradicate in Cornwall and elsewhere in Scilly. Even in the eighteenth century exotic plants were grown in Scilly and easily obtained by close contact with shipping, and doubtless this is the explanation of how it got planted on St Agnes.

VERBENACEAE
Verbena L.

V. officinalis L.   Vervain.   Native.

Roadsides and waste places.   Rare.

M.-.T.-.-. : -.-.-.-.-.   First record: Millett, 1852.

St Mary's: near Tolman Point, Herb. D.-S.; side of road to Peninnis Head, Polunin; Hugh Town, Grose; near Star Castle; about Maypole. Tresco: near Tresco Abbey, Herb. D.-S.

LABIATAE
Mentha L.

M. pulegium L.   Pennyroyal.   Native.

Probably by small ponds.   Very rare.

M.-.-.-.-. : -.-.-.-.-.   First record: Ralfs, 1879.

St Mary's: near the Giant's Grave and at Holy Vale, Ralfs in litt.

to Townsend, 1876, and in his *Flora;* Holy Vale, *c* 1915, Mrs
C. Sandwith.

M. aquatica L.   Water Mint.   Native.
Marshes, by pools, and wet roadsides.   Rare.
M.-.T.-.-. : -.-.-.-.-.   First record: Townsend, 1864.
St Mary's: marsh at Old Town, Townsend; Higher Moors, Herb.
D.-S.; Porthellick; wet roadside near Maypole. Tresco: Tresco
Pool, Herb. D.-S.

M. spicata L.   Spear Mint.   Established alien.
Roadsides near houses.   Rare.
M.A.-.B.MN. : -.-.-.-.-.   First record: Townsend, 1864.
St Mary's: Holy Vale, C. E. Salmon in Davey, 1909; London,
Grose; Carn Friars. St Agnes: near the coastguard station, 1969,
R. Lancaster. Bryher: probably an escape, Townsend. St Martin's:
Lower Town.

M. longifolia (L.) Huds.   Horse Mint.   Established alien.
Lanesides, damp fields, quarries.   Frequent.
M.A.T.B.MN. : -.-.-.-.-.   First record: Ralfs, 1879.
In Higher and Lower Moors and by cliff paths away from houses
this often looks very 'wild', but in all cases it probably originated
as a garden outcast.

M. x villosa Huds. var. *alopecuroides* (Hull) Briq. (*M. niliaca* auct.).
Established alien.
Garden outcast, increasing.
M.-.-.-.MN. : -.-.-.-.-.   First record: Lousley, 1967.
St Mary's: Pasture north-east of Telegraph, 1963, W. H. Hardaker;
Bar Point, 1962, R. C. L. Howitt; Porthellick Pool, 1970; Low
Pool, 1970. St Martin's: Middle Town, 1952.

Lycopus L.
L. europaeus L.   Gipsywort.   Native.
In the marshes.   Very local.
M.-.-.-. : -.-.-.-.   First record: Millett, 1852.
St Mary's: Old Town Marshes, Townsend; Higher Moors, especially
about Porthellick Pool.

Origanum L.

O. vulgare L. Marjoram. Recorded in Davey, 1909 on the authority of Boyden, but not included in Boyden's own list in Tonkin and Row. Probably an error of transcription by Davey.

Thymus L.

T. drucei Ronn.   Wild Thyme.   Native.
On sand and shallow soils over granite, on walls.   Rare.
M.A.-.-.-.   :   -.-.-.-.E.   First record: Millett, 1852.
St Mary's: 'On wall Sandy Bank, St Mary's *only*', Herb. D.-S.;
hill above the Church, Hugh Town; Tremelethen, Grose. St Agnes:
on sand and rocks at both ends of Gugh Bar (390). Eastern Isles:
Great Ganilly, 1966, P. Z. MacKenzie.

Calamintha Mill.

C. ascendens Jord.   Common Calamint.   Native.
Dry hedgebanks.   Rare.
M.A.-.-.-.   :   -.-.-.-.   First record: Somerville, 1893.
St Mary's: 'Growing in the grass-grown dry "dyke" of a farm garden', 1890, Somerville, Herb. Univ. Glasgow.; Upper Trenoweth Lane (284). St Agnes: Between Turk's Head and Middle Town.

Acinos Mill

A. arvensis (Lam.) Dandy. This species of calcareous soils was included in the list of Boyden in Tonkin and Row, 1893. Probably an error.

Melissa L.

M. officinalis L. Balm. St Agnes: laneside near houses, 1956. No doubt an 'escape' from a garden as is the usual case.

Salvia L.

S. pratensis L. Error. Recorded by Boyden in *Trans. Penzance Nat. Hist. Soc.*, **1889/90**, 186, but Davey, 1909 states that Arthur Bennett examined Boyden's specimen and found it to be *S. horminoides*.

S. horminoides Pourr. (*S. verbenaca* auct.).   Wild Sage.   Native.
Dry banks.   Rare.
M.A.-.-.-.   :   -.-.-.-.   First record: Bree, 1821.
St Mary's: Rare, near Hugh Town, Townsend; near Hugh Town, 1952, Grose; near Lifeboat Station, 1967, F. T. Daniels. St Agnes:

bank near Higher Town; Middle Town, by the Reading Room, 1947, G. Grigson; 1966, H. M. Quick.

Prunella L.
  P. vulgaris L.   Selfheal.   Native.
  Roadsides, pastures, etc.   Frequent.
  M.A.T.B.MN. : S.-.-.-.-.   First record: Millett, 1852.

Stachys L.
  S. arvensis (L.) L.   Corn Woundwort.   Native.
  Weed in cultivated fields.   Local.
  M.A.T.B.MN. : -.-.-.-.-.   First record: Townsend, 1864.
  St Mary's: near Deep Point, Grose; near Porthellick; Pelistry Bay; Carn Friars; Porthloo, B. M. C. Morgan. St Agnes: Ribbons and Wanstall, 1948; Lower Town. Tresco: Herb. D.-S.; New Grimsby, Grose; Old Grimsby; Abbey Gardens; Dolphin Town. Bryher: bulbfields near The Town and Church. St Martin's: Middle Town, Higher Town and below School.

  S. palustris L.   Marsh Woundwort.   Native.
  Usually as a weed in cultivated land, but also damp roadsides and marshes.   Uncommon.
  M.-.T.B.MN. : -.-.-.-.-.   First record: Millett, 1852.
  St Mary's: Millett ex Davey, 1909; cultivated field in centre of the island, Townsend *MS.* 1862; Rare, Townsend, 1864; Ralfs' *Flora;* Wanstall and Ribbons, 1948. Tresco: Ralfs' *Flora;* near Dolphin Town, Herb. D.-S.; abundant as weed in cultivated fields near Dolphin Town; weed in Abbey Gardens; Old Grimsby.Bryher: Churchyard, 1953, Polunin. St Martin's: St Martin's Bay, Grose; roadside near Higher Town.

  S. sylvatica L.   Hedge Woundwort.   Native.
  Shady places.   Rare.
  M.-.T.-.-. : -.-.-.-.-.   First record: Millett, 1852.
  St Mary's: Ralfs' *Flora.* Tresco: in the withy bed near the Abbey. In my experience the usual habitats of *S. palustris* and *S. sylvatica* in Scilly differ from those usual on the mainland. Thus *palustris* occurs in dry cultivated fields, while *sylvatica* inhabits a wet withy bed!

Ballota L.

B. nigra L. subsp. *foetida* Hayek   Black Horehound.   Native.
Roadsides and waste places.   Local.
M.-.-.-.-. : -.-.-.-.-.   First record: Millett, 1852.
St Mary's: Hugh Town, Townsend, 1864; Star Castle Hill and
Old Town Bay, Grose; Old Town, Polunin; Hugh Town; Porth-
mellin; Porth Cressa.
Black Horehound is common round Hugh Town but apparently
absent from the rest of Scilly. It may be an ancient introduction.

Lamium L.

L. amplexicaule L.   Henbit.   Native.
Cultivated field.   Very rare.
-.-.-.-.MN. : -.-.-.-.-.   First record: Lousley, 1967.
St Martin's: One plant in field near Higher Town, Mrs B. H. S.
Russell, April 1960.

L. hybridum Vill.   Cut-leaved Deadnettle.   Native.
Cultivated land and roadsides on sandy soil.   Frequent.
M.A.T.B.MN. : -.-.-.-.-.   First record: Townsend, 1864.

L. purpureum L.   Red Deadnettle.   Native.
Cultivated land and waste places.   Rare.
M.A.T.B.MN. : -.-.-.-.-.   First record: Townsend, 1864.
St Mary's: Star Castle Hill and near Bar Point, Grose. St Agnes:
1960, E. G. Philp. Tresco: Abbey Grounds, Grose. Bryher:
occasional. St Martin's: below Middle Town, scarce.
Plants occur on St Agnes (152) and St Martin's (153) with leaves
somewhat incised and intermediate between this species and
L. hybridum.

L. album L.   White Deadnettle.   Colonist.
Very rare.
M.-.-.-.-. : -.-.-.-.-.   First record: Ralfs, 1879.
St Mary's: Ralfs' *Flora*. The only record.

Galeopsis L.

G. angustifolia Ehrh. ex Hoffm. was recorded as *G. ladanum* from
Tresco by Lawson, 1870. The records in Ralfs' *Flora* and Davey, 1909
were probably based on this, and the species cannot be accepted without
confirmation.

G. tetrahit L. sensu lato. Common Hempnettle. Native.
Very rare.
M.-.-.-.-. : -.-.-.-.-. First record: Lousley, 1967.
St Mary's: Bar Point, 1939, Grose.

Glechoma L.
G. hederacea L. Ground-ivy. Native.
On cliffs under bracken, also gardens, fields and hedges.
Common.
M.A.T.B.MN. : S.-.H.TN.E. First record: Millett, 1852.

Marrubium L.
M. vulgare L. White Horehound. Was recorded by Millett, 1852, without locality. Ralfs searched for it without success (in litt. to Townsend) but repeated the record in his *Flora*. In spite of the exceptional accuracy of the Misses Millett, the species cannot be accepted without confirmation.

Scutellaria L.
S. galericulata L. Greater Skullcap. Native.
On a stony shore and by a small pool. Very local.
-.-.-.-.-. : S.-.-.-.-. First record: Townsend, 1864.
Samson: Stony shore, Townsend; still plentiful on the south-east shore (380).

Teucrium L.
T. scorodonia L. Woodsage. Native.
Heaths and cliff slopes, often under Bracken. Frequent.
M.A.T.B.MN. : S.-.H.TN.E. First record: Townsend, 1864.

Ajuga L.
A. reptans L. Bugle. Native.
Very rare.
M.-.-.-.-. : -.-.-.-.-. First record: Tellam in Davey, 1909.
St Mary's: Old Town, Herb. D.-S.—Miss Anne Dorrien-Smith told me about 1940 that it was still there. In 1959 I noticed a form with reddish leaves in the garden of Tregarthen's Hotel but this was no doubt originally planted.

PLANTAGINACEAE

Plantago L.
P. major L. Greater Plantain. Native.

Tracks and roadsides.   Common.
M.A.T.B.MN. : -.-.-.-.-.   First record: Townsend, 1864.
var. *intermedia* (Gilib.) Syme was recorded by Somerville, 1893.
P. media L. Hoary Plaintain. Has been reported from cultivated land on
Bryher, and from St Agnes, but probably in error.

P. lanceolata L.   Narrow-leaved Plaintain.   Native.
Pastures, dunes, cultivated land and waste places.   Very common.
M.A.T.B.MN. : S.-.H.TN.-.   First record: Millett, 1852.
var. *timbali* (Jord.) Druce was recorded for Scilly by Tellam in
Davey, 1909. If by this is intended a luxuriant plant with acute,
scarious silvery bracts, I have seen it in bulbfields near London,
St Mary's, and near the Church on Bryher. It may be of alien
origin. Cardew and Baker have suggested that *P. timbali* of Jordan
is a different plant (*Rep. Watson B.E.C.*, **2**, 356, 1913).

P. maritima L.   Sea Plantain.   Native.
Sea-cliff.   Very rare.
M.-.-.-.-. : -.-.-.-.-.   First record: this *Flora*.
St Mary's: plentiful in a small area of cliff between Pelistry Bay
and Deep Point (897), 1969, R. Lancaster.
Davey, 1909 quotes 'Gerard, 1633' for the first record of this
species in Scilly and Cornwall but this is doubtful. In Gerard's
Herball, second edition edited by Thomas Johnson, pages 425–6,
'Sea Buck-horne Plantaines' are described and it reads: 'The first
and second of these plants are strangers in England; notwith-
standing I have heard say that they grow upon the rocks in Silley,
Garnsey, and the Isle of Man'. The 'first' is a mountain plant, the
'second' a form of *P. coronopus* from sea-washed coast similar to
forms which grow so freely in Scilly and no doubt also in the Isle
of Man and Guernsey. Many botanists have searched for *P.
maritima* in Scilly, and Smith, 1909, comments on its apparent
absence, but it was not until 1969 that it was first found.

P. coronopus L.   Buckshorn Plantain.   Native.
Dry places, and especially in the turf on cliffs subject to salt spray.
Abundant.   M.A.T.B.MN. : S.AT.H.TN.E.
First record: Gerard, 1633 (see above). Also Townsend, 1864.
Very variable. A common form in the most exposed and dry
places has subentire leaves in a rosette only 1.25–1.5 cm in diameter
('var.' *pygmaea* Lange). This is the plant described in Gerard.

P. lagopus L.   Casual.
Cultivated ground, St Agnes, June 1923, H. Downes in Herb. Kew.

Littorella Berg.
L. uniflora (L.) Aschers.   Shoreweed.   Native.
Submerged and also on the shores of pools.   Rare.
-.A.T.-.-. : -.-.-.-.-.   First record: Townsend, 1864.
St Agnes: Pool in Priglis Bay, Townsend. There are specimens in
his herbarium but I have failed to refind it. Tresco: Tresco Pool,
Townsend. It still occurs plentifully in Abbey Pool both in the
barren submerged state (170), and flowering on the sandy margin
(371, 727).

CAMPANULACEAE
Wahlenbergia Schrad.
W. hederacea (L.) Reichb.   Ivy-leaved Bellflower.   Native.
Ditchsides and wet turf in the marshes.   Local.
M.-.-.-.-. : -.-.-.-.-.   First record: Bree, 1831.
St Mary's: Old Town Marshes, Townsend, 1864; Higher Moor,
Herb. D.-S.; abundant with *Hydrocotyle* in wet turf by Porthellick
Pool, 1963 onwards. Also unlocalised records by Millett, Curnow
and Beeby.

Jasione L.
J. montana L.   Sheepsbit.   Native.
Stone hedges and dunes.
M.A.T.-.MN. : -.-.-.TN.-.   First record: Millett, 1852.
Plentiful on St Mary's and St Martin's, rare on Tresco and St Agnes,
and unrecorded for Bryher.

RUBIACEAE
Sherardia L.
S. arvensis L.   Field Madder.   Native.
Cultivated fields, roadsides, rough cliffs and foreshores.
Frequent.
M.A.T.B.MN. : -.-.-.-.-.   First record: Millett, 1852.

Galium L.
G. mollugo L.   Hedge Bedstraw.   Native.
Very rare.
M.-.-.-.MN. : -.-.-.-.-.   First record: Millett, 1852.
St Mary's: Ralfs' *Flora*. St Martin's: 1938, M. Knox.

G. verum L.   Lady's Bedstraw.   Native.
Sandy places.   Frequent.
M.A.T.B.MN. : S.-.-.TN.E.   First record: Millett, 1852.

G. saxatile L.   Heath Bedstraw.   Native.
Heathy places.   Frequent.
M.A.T.B.MN. : S.-.H.-.-.   First record: Townsend, 1864.

G. palustre L.   Marsh Bedstraw.   Native.
Marshes and poolsides.   Local
M.-.T.-.-. : -.-.-.-.-.   First record: Townsend, 1864.
St Mary's: Higher Moors (265); Old Town Marsh (650); Lower
Moors near Low Pool (664). Tresco: by Abbey Pool.
Here, as elsewhere, this species is variable, but all the material
I have seen belongs to subsp. *palustre*. Townsend, 1864, gave
'var. *elongatum* Presl' as 'common' and in his herbarium there are
specimens of two forms from St Mary's marshes. Further study
may show that subsp. *elongatum* (C. Presl) Lange occurs in the
reed-swamps.

G. aparine L.   Cleavers.   Native.
Hedgebanks, cultivated fields, and shores.   Very common.
M.A.T.B.MN. : S.AT.H.TN.E.   First record: Millett, 1852.
G. spurium L. Error. Recorded in Davey, 1902 and rejected in Davey,
1904.

Rubia L.
R. peregrina L.   Madder.   Native.
Rough cliff slopes, often in brambles.   Very common.
M.A.T.B.MN. : S.-.H.TN.E.   First record: Millett, 1852.

Coprosma J. R. and G. Forst.
C. repens A. Rich (*C. baueri* auct. non Endl.)   Alien.
St Mary's: Planted round the Park, Hugh Town (826, 840) from
which seedlings appear in the gutters of the main street, 1963.
St Agnes: seedlings seen, 1967, D. M. Frowde. Tresco: seedlings
plentiful in wood above Tresco Abbey Garden, 1963. This New
Zealand shrub is already out of cultivation from seed, and is likely
to spread. It is increasingly planted as a windbreak.

CAPRIFOLIACEAE

Sambucus L.

S. nigra L.   Elder.   Native.

Roadsides and rough places, often near buildings.   Frequent.

M.A.T.B.MN. : S.-.-.-.-.   First record: Leland, *c* 1535–1543 (published 1710) . . . 'Innisschawe (ie Tresco) . . . bereth stynkkyng elders'.

Elders occur as isolated bushes scattered through the inhabited islands, and there are two bushes near the ruins on Samson. There is some evidence that it has been planted, and Townsend, 1862 thought it 'probably introduced'. In Cornwall it is regarded as a protection against witches (Davey, 1909), and the late F. Rilstone told me that a generation ago many people dried elder flowers for winter use as a medicine for colds ('elder tea'). It may have been planted by the houses on Samson for one or both these reasons. At Polperro it is used for hedges of the small cliff gardens, and Scillonians in their desperate need for windbreaks are likely to have used elder for this purpose.

Lonicera L.

L. periclymenum L.   Common Honeysuckle.   Native.
Cliff slopes, heaths and walls.   Abundant.
M.A.T.B.MN. : S.-.H.TN.E.   First record: Millett, 1852.

On the mainland we are accustomed to regard Honeysuckle as a climber, but in Scilly, where there are few suitable trees, it sometimes scrambles over brambles and gorse, but is often prostrate on the cliff slopes. Few visitors to Scilly in June can fail to be impressed with the flowers which seem larger, a richer colour, and more sweet smelling than those at home. In places much exposed to the gales, the leaves are exceptionally fleshy—as on rocks on top of South Hill, Samson (715).

L. caprifolium L.   Perfoliate Honeysuckle.   Garden escape.
Tresco: outside Abbey Gardens, 1964, Mary Briggs.

VALERIANACEAE

Valerianella Mill.

V. locusta (L.) Betcke   Common Cornsalad.   Native.
Cultivated ground, roadsides, walls, and sandy shores.   Common.
M.A.T.B.MN. : -.-.-.TN.-.   First record: Ralfs, 1879 (and in litt. to Townsend, 1876).

Ralfs knew this only from St Mary's and says 'not common', and from the lack of early records it seems likely that this species has increased greatly since the introduction of the bulb industry. Plants with somewhat hairy fruits were found on St Mary's by Miss E. S. Todd in 1933 (as *V. olitoria* var. *lasiocarpa* Reichb.), and by me on St Martin's (142).

subsp. *dunensis* (D. E. Allen) P. D. Sell. This differs from the typical subspecies in being much dwarfer, more compact, acaulescent and with leaves only up to 3.5 cm long. It occurs on consolidated dune at West Porth, Tean (618) flowering in April. The characters are not impressive but have been tested in cultivation on material from the Isle of Man (Allen, D. E., *Watsonia*, **5**, 45–46, 1961); it has a western type of distribution, and occurs on Herm in the Channel Isles, and probably in Brittany.

V. carinata Lois.   Keel-fruited Cornsalad.   Colonist.
Bulbfield weed, and roadside.   Very rare.
M.A.-.-.-. : -.-.-.-.-.   First record: Lousley, 1940.
St Mary's: bulbfield near London (139); flower field near Old Town Church, 1952, A. Conolly. St Agnes: dry ditch by road at Higher Town, 1948, P. J. Wanstall.

V. dentata (L.) Poll.   Tooth-fruited Cornsalad.   Colonist.
Very rare.
-.-.T.-.-. : -.-.-.-.-.   First record: Lousley, 1967.
Tresco: 1966, det. Cambridge as var. *mixta*, M. McCallum Webster (10,446), Herb. Lousley.

Centranthus DC.
C. ruber (L.) DC.   Red Valerian.   Naturalised alien.
Walls and disused quarry.   Local.
M.-.-.-.-. : -.-.-.-.-.   First record: Lousley, 1967.
St Mary's: roadside wall, Porthloo, 1953, Polunin; Porthloo, spreading from garden to quarry and walls, 1952; walls near Tregarthen's Hotel, 1970, R. M. Burton.

DIPSACACEAE

Dipsacus L.
D. fullonum L. subsp. *fullonum*   Common Teasel.   Alien.
Upper shore.   Very rare.
M.-.-.-.-. : -.-.-.-.-.   First record: this *Flora*.

St Mary's: Shore at Porth Minick = several plants near Old Town, 1967, P. Z. MacKenzie; about Old Town, 1969, R. Lancaster, 1970. Although regarded as native on the mainland, this recent appearance in Scilly is at a place until recently used as a refuse tip. Probably introduced with cage-bird seed.

Knautia L.
K. arvensis (L.) Coult.   Field Scabious.   Native.
Extinct.
M.-.-.-.-. : -.-.-.-.-.   First record: Millett, 1852.
St Mary's: Not common, Ralfs' *Flora*.
In view of the abundance of this species in Cornwall, and the reliability of the Misses Millett, these records must be accepted, but no recent observer has reported it.

Succisa Haller
S. pratensis Moench Devil's-bit Scabious. Error? Recorded by Tellam in Davey, 1909 and by Boyden in Tonkin and Row, 1893. There are suitable habitats, but the late Mr F. Rilstone wrote that Cornish botanists who should know better confused this with *Jasione montana*, partly on account of the similarity between the names 'Devils-bit' and 'Sheepsbit'. No recent botanist has been able to find it.

COMPOSITAE
Senecio L.
S. jacobaea L.   Common Ragwort.   Native.
Roadsides, rough pastures, and sandy ground.   Common.
M.A.T.B.MN. : S.AT.-.TN.E.   First record: Millett, 1852.

S. sylvaticus L.   Heath Groundsel.   Native.
Heathy and sandy ground.   Common.
M.A.T.B.MN. : S.AT.-.TN.E.   First record: Townsend, 1864.
On rocky islets, such as Nor-Nour and Great Innisvouls (325) this species grows under heavy salt concentrations.

S. vulgaris L.   Common Groundsel.   Native.
Cultivated grounds, walls, roadsides, and dunes.   Very common.
M.A.T.B.MN. : S.AT.H.TN.E.   First record: Borlase, 1756.
Sometimes with woolly stems on dunes, as, for example, at Bar Point, St Mary's (601).

S. mikanioides Otto ex Walp.   German Ivy   Naturalised alien.
Walls, banks, on boulders on shore, in elm copse, etc.   Common.
(Fig 11, p 251).   M.A.T.-.-. : -.-.-.-.-.   First record: Downes
in *Rep. B.E.C.*, **7**, 1052, 1926.
St Mary's: Hill above Church, Hugh Town (229); Porth Mellon;
Rocky Hill; dunes, Old Town, Polunin; Old Town churchyard and
spread from there into elm copse. St Agnes: near Parsonage (385);
near Lighthouse, 1952, Grose; field below Parsonage; Priglis Bay,
1952, naturalised in several places; abundant by chapel, 1952;
strand north of Middle Town, 1953, Polunin; on boulders of
storm beach, Porth Killier, 1957. Tresco: abundantly naturalised
about Abbey (238); Gimble Porth; under Monument Hill; near
Abbey Farm; Appletree Banks.
German Ivy flowers in November and puzzles many visiting
botanists who find only the leaves. It is thoroughly established and
sometimes a long way from houses. I first noticed it in 1939 and it
must have been established for many years before then. In the
Channel Isles it was found at St Catherines, Jersey by N. D. Simp-
son in 1923 and is still plentiful there under winter climatic conditions less favourable than it enjoys in Scilly.

Tussilago L.
T. farfara L.   Coltsfoot.   Native.
On a roadside.   Very rare.
M.-.T.-.-. : -.-.-.-.-.   First record: Lousley, 1967. (but see below)
St Mary's: In flower, Rev. Wm. T. Price in letter to *The Times*,
December 12, 1931. Tresco: In flower on January 6, 1881, *Gdnrs'
Chron.*, **1881,** 84; in one spot by Pool Road near Great Pool, 1939,
Grose and Lousley.

Petasites Mill.
P. fragrans (Vill.) C. Presl   Winter Heliotrope.   Established
alien.
Roadsides and fields.   Local.
M.A.T.B.-. : -.-.-.-.-.   First record: Lousley, 1967.
St Mary's: Holy Vale, 1939, Grose; near Old Town Church, 1952,
Grose and Lousley; side of road to Peninnis Head, 1953, Polunin;
near Pelistry. St Agnes: 1948, Ribbons and Wanstall; near Light-
house, 1952, Grose. Tresco: thoroughly established in and around
the Abbey Gardens, 1953. Bryher: well established in two fields,
Northward, 1953.

Brachyglottis J. R. and G. Forst.

B. repanda J. R. and G. Forst. (Fig 9.) This New Zealand tree, with its large leaves felted below, has been extensively planted as a windbreak in St Mary's, St Agnes, Tresco and St Martin's. It flowers and fruits freely in April but, although it is sometimes found in places not obviously planted, I have no evidence that it spreads by seed.

Gazania Gaertn.

G. x splendens Lemaire   Established alien.
St Agnes: 1960, Howitt. St Martin's: On cliff face near Yellow Rock below Middle Town, 1956 (738), 1957 (810), onwards. It grows here in a situation where it would be difficult to plant, and could hardly arise directly from refuse, and it is not near houses.

Calendula L.

C. officinalis L.   Garden Marigold.   Naturalised alien.
Sandy and rocky shores, walls, roadsides, waste places, etc. Common.
M.A.T.B.MN. : -.-.-.-.-.   First record: Lousley, 1967.
Garden Marigold is thoroughly naturalised in Scilly although never very far from houses. I first noticed it in 1939 but it was not until 1952 that I realised its permanence, and the present populations are essentially the result of increase over the last 20 years.

C. arvensis L.   Field Marigold.   Casual.
The only record—St Mary's: weed in bulbfield, 1960, Howitt.

Inula L.

I. helenium L.   Elecampane.   Denizen.
In a field.   Probably extinct.
M.-.-.-.-. : -.-.-.-.-.   First record: Bree in Forbes, 1821.
St Mary's: Bree, as above; Millett, 1852; Curnow, 1876; field near Old Town Marsh, Ralfs in litt. to Townsend, 1876.
The locality was near to Castle Ennor, the centre of government until Elizabethan times, and Elecampane is likely to have been planted for medicinal purposes.

Pulicaria Gaertn.

P. dysenterica (L.) Bernh.   Common Fleabane.   Native.
In a marsh.   Very rare.
M.-.-.-.-. : -.-.-.-.-.   First record: Millett, 1852.

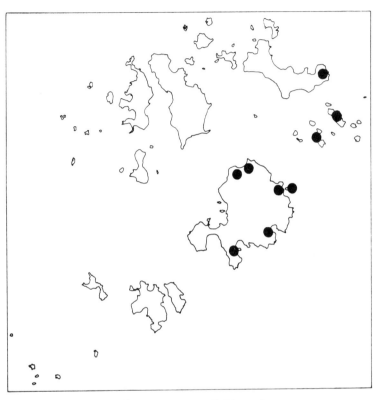

Solidago virgaurea, Goldenrod

St Mary's: Common, Ralfs' *Flora*—a statement of frequency which is clearly untrue; top of Higher Moors Valley below Longstone Farm, 1967, P. Z. MacKenzie (865). There is also an unlocalised specimen in Herb. Dorrien-Smith.

Filago L.

F. vulgaris Lam. (*F. germanica* L. non Huds.).   Common Cudweed.   Native.

Roadsides and cultivated fields.   Local.

M.S.-.-.-. : -.-.-.-.   First record: Millett, 1852.

St Mary's: Bar Point, Grose; roadside, Carn Friars; frequent about Maypole; Holy Vale (301); Porthellick, Grose; Pelistry Lane; near Telegraph. St Agnes: uncommon, H. M. Quick.

F. pyramidata L. (*F. spathulata* C. Presl.) Recorded by Somerville, 1893 but certainly an error.

F. minima (Sm.) Pers. Recorded by Boyden in Tonkin and Row, and repeated in Davey, 1909. Confirmation required.

Gnaphalium L.

G. uliginosum L.   Marsh Cudweed.   Native.
Ditch and pool sides, and damp places in fields and tracks.
Frequent.
M.A.T.B.MN. : S.-.-.-.-.   First record: Townsend, 1864.

Solidago L.

S. virgaurea L.   Goldenrod.   Native.
Heathy ground near the coast.   Rare. (Map 11, p 244.)
M.-.-.-.MN. : -.-.-.-.E.   First record: Ralfs' *Flora*, 1879.
St Mary's: Not common, Ralfs in litt. to Townsend, 1876; by road to Pelistry Bay; Toll's Island; Porthellick; Bar Point; near Bant's Carn. St Martin's: cliffs around Chapel Down. Eastern Isles: Middle Arthur, Herb. D.-S.; Great Ganilly.
It is curious that this common Cornish plant should have such a restricted distribution in Scilly where there is so much apparently suitable ground. It extends over a sharply defined area from the north and east coasts of St Mary's, through the Eastern Isles to St Martin's Head.

Aster L.

Aster tripolium L. was recorded from Tresco by Ralfs in his *Flora*, 1879 but probably in error.

Erigeron L.

E. mucronatus DC.   Mexican Fleabane.   Naturalised alien.
M.-.T.-.MN. : -.-.-.-.   First record: Lousley, 1967.
St Mary's: Old Town, on walls of ruined house, 1953, Polunin. Tresco: walls near the Abbey, 1957. St Martin's: walls, Higher Town, 1957.
During the last ten years it has spread considerably on Tresco and St Martin's.

Bellis L.

B. perennis L.   Daisy.   Native.
Pastures and maritime turf, mainly on ground which is trodden or
grazed.   Common.
M.A.T.B.MN. : S.-.H.TN.-.   First record: Millett, 1852.
Townsend, 1862, marked this as 'local' and said 'in many large
tracts it would be difficult to meet with a specimen'. This is true but
the daisy is fairly evenly distributed over the islands. A large patch
with all the flowers discoid was found on Samson Hill, Bryher by
D. McClintock in April 1953 (559). In the bulbfields there is a
luxuriant form with the stem much branched below and peduncles
15–18 cm long—e.g. By Borough Road, Tresco (779).

Olearia Moench
O. traversii F.v.M., from Chatham Island, is now widely used as a
first front hedging plant, especially on Tresco. Small plants thought
to be self-sown are reported by R. C. L. Howitt.

Anthemis L.
A. cotula L.   Stinking Chamomile.   Native?
Cultivated and waste ground.   Uncommon.
M.A.-.B.-. : -.-.-.-.   First record: Townsend, 1864.
St Mary's: Holy Vale, Townsend; weed in bulbfield, Halangy (400).
St Agnes: H. M. Quick. Bryher: waste ground, Northward (339).

Chamaemelum Mill.
C. nobile (L.) All.   Common Chamomile.   Native.
Greens, turfy pastures, and downs.   Very common.
M.A.T.B.MN. : -.-.-.-.   First record: Borlase, 1756.
The abundance of this plant is a feature of Scilly in late summer
when the white flowers colour tracks and the scent from the foliage
delights the walker.

Achillea L.
A. millefolium L.   Yarrow.   Native.
Roadsides, walls and sandy turf.   Common.
M.A.T.B.MN. : S.-.H.TN.E.   First record: Townsend, 1864.
A very handsome form with deep purplish-rose flowers occurs on
the sandy bar of Great Ganilly (321), and on Tean (f. *purpurea*
(Gouan) Schinz and Thellung).

Otanthus Hoffmanns. and Link
O. maritimus (L.) Hoffmanns. and Link   Sea Cottonweed.
Native.
Sand-dunes in *Ammophila*.   Extinct. (Plate p 225.)
-.-.-.-.MN. : -.-.-.-.-.   First record: Hosking in *Rep. B.E.C.*, **5,** 490, 1917.
St Martin's: Hosking recorded this as 'Land's End district', and Thurston stated that it was seen in the same locality by the Rev. F. Granville May and the Rev. (afterwards Bishop) J. Hannington in 1877 (Thurston and Vigurs, 1922/A, 73). That the locality was the north-east coast of St Martin's was revealed by W. B. Haley in 1921 (Haley, 1921), and the following year H. Downes reported 'Several hundred plants in one of the Scilly Isles' and revealed that this was the locality of Hosking, with whom he was in correspondence (*Rep. B.E.C.*, **6,** 385, 1922).
The late Mr Spencer Savage told me that in 1925 the patch was about 30 yards long, and he gave me a series of photographs taken then. Sir Edward Salisbury has lantern slides showing it in quantity at about the same time, but by the time I saw *Otanthus* in September 1936 it was greatly reduced, looked weak, and produced no flowers. By 1938 there was no trace of the plant.
Sea Cottonweed is abundant on the shores of the Mediterranean and has made repeated attempts to colonise Britain, which is on the extreme northern limit of its range. The only colony which at present persists is in Wexford: the others have flourished for a time and then disappeared. In Jersey, where I saw it in 1926, it similarly weakened before it was lost. The 60 years it grew in Scilly coincided with a period during which there was no frost severe enough to damage the Cordylines ('palms') on the mainland of Cornwall, and then, as the late F. Rilstone pointed out, came two severe winters which killed off many inland specimens and made an almost clean sweep of the tenderer shrubby Veronicas. It seems likely that the main cause of the extinction of *Otanthus* was climatic. The end may have been hastened by a grazing animal tethered on the spot, or by the dumping of a load of seaweed (*The Scillonian,* **25,** 185, 1950).

Tripleurospermum Schultz Bip.
T. maritimum (L.) Koch subsp. *inodorum* (L.) Hyland.   Scentless Mayweed.   Native.

Shores, waste and cultivated ground and roadsides.   Very common.
M.A.T.B.MN. : S.AT.H.TN.E.   First record: Millett, 1852.
On the rocky shores of small islands this species is abundant with fleshy leaves and larger flowers than usual. Such plants are now treated as var. *salinum* (Wallr.) Clapham. They were recorded by Townsend, 1864 as *Matricaria maritima* L. which is a more northern plant now treated as a subspecies.
Plants with double flowers (*fl. pleno*) are frequent on ground rich with the guano of sea-birds as on Annet (820), Great Ganinick (328) and Middle Arthur (712).

Matricaria L.
M. matricarioides (Less.) Porter   Pineapple Weed.
Naturalised alien.
Field gateways, cultivated and waste land.   Local.
M.-.T.-.MN. : -.-.-.-.   First record: Lousley, 1967.
St Mary's: Star Castle Hill and Holy Vale, 1939, Grose; Porthloo, 1952; Carn Friars, 1956. Tresco: New and Old Grimsby (spec.), 1968, D. A. Cadbury. St Martin's: Noted in 1938 by Dallas; Middle Town and two places at Higher Town, 1939, Grose; Lower Town, 1940; by track from New Quay, 1940.
Davey, 1909 records the rapid spread of this species on the mainland of Cornwall in the 12 years he had known it. In Scilly it was not noticed until 1938, and it is still known only from three islands. It spread most rapidly on St Martin's and is now fast extending its range on St Mary's.

Chrysanthemum L.
C. segetum L.   Bothum, Bothams.   Corn Marigold.   Native?
Weed in arable fields.   Abundant.
M.A.T.B.MN. : -.-.-.-.   First record: Townsend, 1864.
Many of the bulbfields in May and June are golden with the blooms of this species, making a glorious pattern of colour when seen from the air. In Scilly it is often a magnificent plant, two feet tall, much branched, with up to 30 flowers from one root. Townsend apparently did not find it in the quantity in which it now occurs, which supports the theory that it is a Mediterranean species introduced into this country and still increasing, but it is also likely that the bulbfields favour it more than the old mixed economy of corn and potatoes. The local name 'Bothum', or 'Bothams' as I have heard it called

on Tresco and St Agnes, has been in use for a century (Scott and Rivington, 1870). The present abundance is such that attempts have been made to market the flowers.

C. leucanthemum L. (*Leucanthemum vulgare* Lam.)   Ox-eye Daisy.   Native.
Meadows, walls, roadsides and churchyards.   Local.
M.-.T.B.MN. : -.-.-.-.-.   First record: Millett, 1852.
St Mary's: roadside near Telegraph; wall in Hugh Town; abundant in two meadows near Tremelethen; abundant in field off High Cross Lane, and one near Salakee. Tresco: in the churchyard. Bryher: Samson Hill, Grose; in the churchyard. St Martin's: near Brandy Point, rare.

C. parthenium (L.) Bernh. (*Tanacetum parthenium* (L.) Schult-Bip.)
Feverfew.   Established alien.
Recorded by Millett, 1852, and by Somerville, 1893 as 'introduced', this species was observed on St Mary's by R. C. L. Howitt in 1956. The locality is likely to be walls near houses, but I have no further information.

C. vulgare (L.) Bernh. (*Tanacetum vulgare* L.)   Tansy.
Denizen.
A laneside.   Very rare.
-.-.T.-.-. : -.-.-.-.-.   First record: Borlase, 1756, 78.
Tresco: Lawson, 1870; Back Lane.
Borlase said it was wild but gave no locality, and it was included in the list of the Misses Millett, 1852. Tansy was formerly much cultivated as a pot-herb and for medicinal purposes, and it was probably introduced to Tresco by the monks.

Helichrysum Mill. em. Pers.
H. petiolatum (L.) DC. This native of South Africa grows on the sand-dunes at Pentle Bay, Tresco (517, det. D. Killick). It was almost certainly planted there in common with many other aliens, and the reason for mentioning it here is that it has on several occasions been reported as *Otanthus maritimus*. Also in a pit above Tremelethen, St Mary's, 1969, R. Lancaster.

Artemisia L.
A. vulgaris L.   Mugwort.   Native.
Roadsides and waste places.   Rare.

M.-.-.B.-.. : -.-.-.-.-.   First record: Millett, 1852.
St Mary's: Star Castle, Grose; Hugh Town, Grose; Porthcressa; The Park; Tolman Point. Bryher: The Town; near Post Office.

A. absinthium L. Wormwood.   Native.
Roadsides and waste places.   Uncommon.
M.A.T.-.MN. : -.-.-.-.-.   First record: Millett, 1852.
St Mary's: About Heugh Town, Townsend, 1864; near Old Town. St Agnes: common about the lighthouse and Higher Town. Tresco: New Grimsby. St Martin's: Middle Town, Grose.
A. maritima L. Recorded from 'Scilly Isles, Somerville sp.', A. Bennett in *J. Bot.*, **43,** 53, 1905. I have been unable to trace the specimen and the species is unlikely to occur.

Carlina L.
C. vulgaris L.   Carline Thistle.   Native.
Probably extinct.   This was recorded by Millett, 1852, and by Ralfs in litt. to Townsend, in both cases without indication of the locality. It has not been noted by any other botanist but there is no reason to doubt that it occurred.

Arctium L.
A. lappa L. This species was listed by the Misses Millett, 1852 and by Boyden in Tonkin & Row, 1893. In both lists this is the only *Arctium*, and no doubt the following species was the one found.

A. minus Bernh. subsp. *minus*   Common Burdock.   Native.
Roadsides and waste places.   Frequent.
M.A.T.B.MN. : -.S.-.-.-.E.   First record: Townsend, 1864.
Subsp. *nemorosum* (Lejeune) Syme was collected by the edge of the main drive to the Abbey Gardens, Tresco, in 1969 by R. Lancaster. The material was determined by Dr F. Perring who remarked that it was the first material of the subspecies he had received from south-west England.
The common *Arctium* of Scilly appears to be *A. minus* subsp. *minus*

Fig 11 Established Aliens: 1 *Oxalis megalorrhiza* (calyx x3, part of under-surface of leaf x1½). 2 *Oxalis pes-caprae* (part of undersurface of leaf x1). 3 *Oxalis articulata*. 4 *Agapanthus praecox*. 5 *Senecio mikanioides*

2

5

3

4

and subsp. *nemorosum* is known only from the drive to Abbey Gardens. It is not possible to assign the old records of *A. pubens* Bab. and *A. vulgare* (Hill) Evans to the subspecies as now understood. On the uninhabited islands Burdock grows about the ruined houses on Samson, and on Great Ganilly; in both cases probably introduced on human clothing.

Carduus L.
  C. tenuiflorus Curt.   Slender-headed Thistle.   Native.
Waste places, roadsides and rocky shores.   Very common.
M.A.T.B.MN. : -.AT.-.TN.E.   First record: Townsend, 1864.
C. nutans L. was reported from St Mary's by Ralfs in litt. to Townsend, 1876 (repeated in Davey, 1909). Casual or error.

Cirsium Mill.
  C. vulgare (Savi) Ten.   Spear Thistle.   Native.
Waste ground, roadsides, field borders, rocky shores, etc.
Common.
M.A.T.B.MN. : S.AT.H.TN.E.   First record: Townsend, 1864.

  C. palustre (L.) Scop.   Marsh Thistle.   Native.
Marshes and damp meadows.   Local.
M.-.-.-.-. : -.-.-.-.-.   First record: Townsend, 1864.
St Mary's: Common in the marshes, Townsend; Higher and Lower Moors.

  C. arvense (L.) Scop.   Creeping Thistle.   Native.
Cultivated fields, meadows, waste ground, etc.   Frequent.
M.A.T.B.MN. : -.-.-.-.-.   First record: Townsend, 1864.

Centaurea L.
  C. nigra L. subsp. *nemoralis* (Jord.) Gugler   Hardhead.
Native.
Meadows, downs, roadsides, churchyards.   Rare.
M.A.T.B.MN. : -.-.-.TN.-.   First record: Millett, 1852.
St Mary's: Hugh Town, Grose; hill above Hugh Town church; Golf Links; The Garrison; Porthmellin; meadow near Tremelethen; road to Old Town; Borough; Trewince. St Agnes: E. G. Philp, 1960. Tresco: New Grimsby, Grose; in churchyard; Back Lane. Bryher: in churchyard. St Martin's: Higher Town. Tean. Although

it occurs in all the inhabited islands, this is usually in very small quantity, and it is much rarer than the number of records would suggest.

C. solstitialis L.   St Barnaby's Thistle.   Casual.
Found by J. D. Grose at Trewince, St Mary's in 1939.

Cichorium L.
C. intybus L.   Chicory.   Denizen.
Cultivated and waste ground.   Very rare.
M.A.T.-.MN. : -.-.-.-.   First record: Lousley, 1967.
St Mary's: Trewince, one plant, 1956. St Agnes: 1966, H. M. Quick. Tresco: wild part of a cottage garden, Mrs O. Moyse, *Scillonian*, **27**, 214–216, 1952. St Martin's: Middle Town, Grose, 1939.

Lapsana L.
L. communis L.   Nipplewort.   Native.
Roadsides, field borders and waste places.   Frequent.
M.A.T.B.MN. : -.-.-.-.   First record: Millett, 1852.

Hypochoeris L.
H. radicata L.   Long-rooted Catsear.   Native.
Pastures, hedges, heathy ground.   Common.
M.A.T.B.MN. : S.-.H.TN.E.   First record: Millett, 1852.
Plants in flower in the spring often have more or less glabrous leaves.

Leontodon L.
L. autumnalis L.   Autumn Hawkbit.   Native.
In grassland dominated by *Juncus articulatus*.   Rare.
M.A.-.-.-. : -.-.-.-.   First record: Townsend, 1864.
St Mary's: in the middle of Old Town Marsh, 1953, Polunin. St Agnes: 1948, Ribbons and Wanstall. Townsend gave this as 'common' but the only definite locality known to me is the one discovered by Polunin.

L. hispidus L.   Greater Hawkbit.   Native.
Meadows, and sandy soils where basic.   Rare.
-.A.T.-.MN. : S.-.-.-.E.   First record: Townsend, 1864.
St Agnes: 1948, Ribbons and Wanstall. Tresco: Abbey Grounds. St Martin's: Lower Town and White Island. Samson. Eastern Isles: Nor-Nour, Grose.

L. taraxacoides (Vill.) Mérat   Lesser Hawkbit.   Native.
Sandy places and bare places on coast.   Abundant.
M.A.T.B.MN. : S.-.-.TN.E.   First record: Townsend, 1864.

Picris L.
P. echioides L.   Prickly Oxtongue.   Native.
Roadsides.   Very local.
M.-.-.B.-. : -.-.-.-.-.   First record: Salmon in Davey, *J. Roy. Inst. Cornw.*, **52,** 1906.
St Mary's: Salmon, as above; roads about the Castle and Hugh Town, Smith, 1909; Old Town, Old Town Bay, Grose; still restricted to the area from The Garrison to Hugh Town and Old Town. Bryher: in village, 1955, Howitt.
The failure of the early botanists to record this, and its distribution about Hugh Town, suggest that *P. echioides* may be a recent introduction, but it is accepted as native in Cornwall, where it is also rather a local plant.
P. hieracioides L. Recorded from St Mary's by Lawson in Davey, *J. Roy. Inst. Cornw.*, **51,** 1905, but *P. echioides* was no doubt intended.

Sonchus L.
S. arvensis L.   Corn Sowthistle.   Native.
Cultivated ground, roadsides, marshes, waste places and shores. Local.
M.A.T.B.MN. : -.-.-.-.-.   First record: Lawson, 1870.
On St Mary's and Bryher this species is very common, but rare on the other three inhabited islands. In 1967, Mr P. Z. MacKenzie measured plants 8 ft tall in Lower Moors, St Mary's.

S. oleraceus L.   Common Sowthistle.   Native.
Cultivated ground, roadsides, rocky shores.   Abundant.
M.A.T.B.MN. : S.-.H.TN.E.   First record: Millett, 1852.

S. asper (L.) Hill   Rough Sowthistle.   Native.
Sandhills, arable land, waste places, shores.   Frequent.
M.A.T.B.MN. : S.AT.H.TN.E.   First record: Townsend, 1864.
While *S. oleraceus* is much more common in man-made habitats, *S. asper* has been recorded mainly from sandhills and shores. Both occur on the shores of uninhabited islands but there *S. asper* is slightly more frequent.

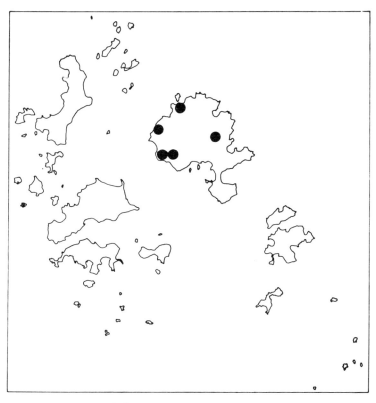

Hieracium umbellatum subsp. *bichlorophyllum*, Hawkweed

Hieracium L.
   H. umbellatum L. subsp. *bichlorophyllum* (Druce and Zahn) Sell and
West.   Native.
In bracken, and by healthy tracks and roadsides.   Local.
(Map 12, p 255).   M.-.-.-.-. : -.-.-.-.-.
First record: Smith, 1909.   First evidence: Ley, 1884—see below.
St Mary's: Smith, 1909; High Cross Lane, Herb. Dallas; by track
from Innisidgen Carn to Bar Point (279, det. Pugsley as *H. umbel-
latum*); Pelistry; Halangy Point (493); near Bant's Carn, in *Pterid-
ium*, Polunin; east shore of St Mary's, 1884, A. Ley det. Sell and
West, Herb. Cantab.

All the material I have seen is the broad-leaved south-western subsp. *bichlorophyllum* and not typical *H. umbellatum* which is more rigid with narrower leaves.

Pilosella Hill

P. officinarum C. H. and F. W. Schultz (*Hieracium pilosella* L.) Mouse-ear Hawkweed.   Native.
Sand-dunes.   Rare.
M.-.T.-.-. : -.-.-.-.-.   First record: Millett, 1852.
St Mary's: Ralfs' *Flora;* Bar Point. Tresco: Herb. D.-S.

Crepis L.

C. vesicaria L. subsp. *taraxacifolia* (Thuill.) Thell.   Beaked Hawksbeard.   Colonist.
Roadsides and field borders.   Local, but increasing.
M.A.-.-.MN. : -.-.-.-.-.   First record: Lousley, 1940.
St Mary's: About Holy Vale, 1939; Porthloo, 1939; fields towards Peninnis, 1940; about Hugh Town, 1940; by 1963 abundant and widespread. St Agnes: near Parsonage, 1957; common as bulbfield weed, 1959. St Martin's: below School, 1953.
When first found in 1939 this species was in quantity but exceedingly local and was probably a very recent arrival. Its spread in St Mary's during the next 20 years was spectacular. In St Agnes it apparently arrived much later but spread equally rapidly.

C. capillaris (L.) Wallr.   Smooth Hawksbeard.   Native.
Roadsides, hedgebanks and arable ground.   Common.
M.A.T.B.MN. : -.-.-.-.-.   First record: Townsend, 1864.
In Scilly this grows throughout the winter, and I have collected it in full flower in April.

Taraxacum Weber

T. officinale Weber.   Common Dandelion.   Native.
Pastures, bulbfields, roadsides, waste places.   Common.
M.A.T.B.MN. : -.-.-.-.-.   First record: Millett, 1852.

T. laevigatum (Willd.) DC. (*T. erythrospermum* Andrz. ex Bess.).
Native.
Consolidated dunes.   Probably common.
-.A.-.B.MN. : -.-.-.TN.-.   First record: Lousley, 1967.
St Agnes: by Gugh Farm. Bryher: Rushy Bay, 1952, A. Conolly. St Martin's: 1957, M. Jaques. Tean: Herb. Dallas.

TS INTRODUCED FROM NEW ZEALAND: (*top*) *Pittosporum crassifolium*, a wind-break shrub, sown on rocks on Bryher; (*bottom*) New Zealand Flax, *Phormium colensoi*, self-sown in t Bay, St Martin's. First recorded in 1920 from cliffs in the background, it is now abundant and has spread to the dunes

Flowering in April and May, this species is certainly heavily under-recorded.

British Taraxaca are currently under review by Dr A. J. Richards who has determined the following specimens:—

T. adamii Claire   St Mary's: Lane to Pelistry Bay (900) R. M. Burton.

T. hematiforme Dt.   St Mary's: Pelistry Bay (7912), Mrs C. Harvey; Porthcressa (905), R. M. Burton. Tresco: sandy turf (913), Mrs C. Harvey.

T. oxoniense Dt.   St Agnes: (847), D. McClintock. Tresco: Appletree Banks (906), R. M. Burton.

MONOCOTYLEDONES
ALISMATACEAE

Baldellia Parl.

B. ranunculoides (L.) Parl.   Lesser Water-plantain.   Native. Shore of a pool.   Very local.

M.-.T.-.-. : -.-.-.-.   First record: Lawson, 1870.

St Mary's: Ralfs' *Flora*. Tresco: Lawson, 1870; edge of little pool, Herb. D.-S.; Margin of Abbey Pool.

Alisma L.

A. plantago-aquatica L. Error. Recorded in Townsend, 1864, but in his 1862 manuscript he wrote 'I am not sure but I think I observed the leaves. Tresco?' He may have mistaken the leaves of *Baldellia* for young plants.

BUTOMACEAE

Butomus L.

B. umbellatus L. Extinct and probably planted. This was recorded by the Misses Millett, 1852, and J. B. Montgomery, 1854—in both cases without locality. In Ralfs' *Flora*, 1879 the locality was given as 'Marsh near Hugh Town' but this may have been just a guess, and Ralfs said he searched for it without success (in litt., 1876). The species was grown in Tresco Abbey Gardens in 1889 (Teague, 1889), and is not known as a native in extreme south-west England.

JUNCAGINACEAE

Triglochin L.

T. palustris L.   Marsh Arrowgrass.   Native. Marshy meadows.   Rare.

M.-.-.-.-. : -.-.-.-. First record: Townsend, 1864.
St Mary's: Old Town Marsh, Towns.; Marsh near Old Town,
Ralfs' *Flora;* High Moors, Herb. D.-S.

ZOSTERACEAE

Zostera L.
Z. marina L.   Common Eel-grass.   Native.
Submerged on the sandy bottom of the shallow sea between the
ring of islands.   Abundant.
M.-.T.B.MN. : S.-.-.-.E.   First record: Woods, 1852.
The distribution of this species is almost continuous from Bar
Point, St Mary's to Samson, Bryher and Tresco, and to St Martin's
and the Eastern Isles and great beds of it may be seen by looking
over the side of a boat at low tide. It seems not to extend across
the deeper water to St Agnes. The widespread reduction in Eel-grass
which set in generally about 1933 appears to have had little effect
in Scilly, where my notes show that it was locally abundant in
1938–1940, but later it was drastically reduced. In 1950, A. A.
Dorrien-Smith reported that there was little left on the flats (i.e.
between the islands) but it was again spreading on sand-banks at
Porth Cressa. (*Scillonian*, **25**, 77, 1950). It must have recovered
rapidly as in 1952 I found it plentiful off Bar Point, St Mary's
(488), between Samson and Bryher (521), at Rushy Bay, Bryher,
and off Great Ganinick in the Eastern Isles.
Tellam recorded *Z. marina* var. *angustifolia* Hornem. (in Davey,
1909), and Woods' plant was narrow-leaved. Earlier I treated these
records as belonging to *Z. angustifolia* (Hornem.) Reichb. (Lousley,
1967) but on reflection I think it more likely that they were narrow-
leaved plants of *Z. marina* such as I have occasionally seen in
Scilly.

POTAMOGETONACEAE

Potamogeton L.
P. natans L.   Floating Pondweed.   Native.
Probably floating in ponds.   Very rare.   M.-.T.-.-. : -.-.-.-.
First record: W. Curnow ex Davey, *J. Roy. Inst. Cornw.*, **52**, 19,
1906.
St Mary's: Higher Moors, Herb. D.-S. teste Edinburgh. Tresco:
W. Curnow teste Drabble ex Davey, op cit.

P. polygonifolius Pourr.   Bog Pondweed.   Native.
Peaty pools and mud.   Very rare.
M.-.T.-.-. : -.-.-.-.-.   First record: Townsend, 1864.
St Mary's: Marshes, Townsend; Boggy part of Higher Moors
(413—teste Dandy and Taylor), still there in very small quantity,
1967, P. Z. MacKenzie (spec.); Town Marsh, 1955, Howitt (spec.);
Porthellick, 1955, Howitt (spec.). Tresco: Ralfs' *Flora;* 1955,
Howitt.

P. perfoliatus L.   Perfoliate Pondweed.   Native.
In a pond.   Very rare.   -.-.T.-.-. : -.-.-.-.-.
First record: Ralfs in litt to Townsend, 1876.
Tresco: Pond, Ralfs, loc cit and *Flora;* Pond near Tresco Abbey,
1877, W. Curnow in 'Scrap-book' at Tresco Abbey.

P. pusillus L.   Lesser Pondweed.   Native.
In large and small pools.   Very rare.   M.-.T.-.-. : -.-.-.-.-.
First record: Ralfs in litt to Townsend, 1876.
St Mary's: Ralfs, loc cit and *Flora;* Pool near Old Telegraph Station,
W. Curnow in 'Scrap-book' at Tresco Abbey. Tresco: Little Pool,
1929, Herb. D.-S. teste Edinburgh. In addition, *P. pusillus* from
Isles of Scilly', July 1890, (*J. Bot.* **31**, 119, 1893) is supported by a
specimen in the Herbarium of the British Museum determined by
Dandy and Taylor (*J. Bot.*, **78**, 5, 1940).

P. pectinatus L.   Fennel-leaved Pondweed.   Native.
In freshwater and brackish pools.   Rare.
M.A.T.-.-. : -.-.-.-.-.   First record: Townsend, 1864.
St Mary's: Marshes, Townsend; marsh near Old Town, Ralfs'
Flora; pool in Old Town Bay, 1876, as 'var. *marinus*', W. Curnow
in 'Scrap-book' at Tresco Abbey. St Agnes, pool, Porth Coose,
1953, Polunin (spec.). Tresco: Big Pool, Herb. D.-S.; Great Pool,
Grose; Abbey Pool (185, teste Dandy and Taylor); The Duckery,
Great Pool (712).

RUPPIACEAE

Ruppia L.
R. maritima L.   Beaked Tassel-pondweed.   Native.
Brackish pools and ditches.   Rare.   M.A.T.B.-. : -.-.-.-.-.
First record: Ralfs in litt to Townsend, 1876.

St Mary's: 1873, W. Curnow in Herb. S. L. B. I.; Ralfs, loc cit; marsh near Old Town, Ralfs' *Flora;* pool in meadow, Old Town, 1876, W. Curnow in 'Scrap-book' at Tresco Abbey. St Agnes: brackish pool, 1877, W. Curnow in 'Scrap-book'; Priglis Pool. Tresco: The Duckery, Great Pool (711). Bryher: Great Pool (729).

LILIACEAE

Phormium J. R. and G. Forst.

P. tenax J. R. and G. Forst. sensu lato.   New Zealand Flax. Naturalised alien.

Cliffs, hedges, sand-dunes, lanesides, etc.   Frequent, and spreading.   M.-.T.B.MN. : -.-.-.-.-.

First record: Hill, 1920 (*P. colensoi*).

St Mary's: cliffs near Watermills. Tresco: Appletree Banks; Monument Hill; by Pool Road, etc. Bryher: established in hedge-banks in several places. St Martin's: see *P. colensoi*, below.

P. tenax has a long history as a garden plant at Tresco Abbey. By 1872 it was remarked that it grew all the year round in the open without protection (*Gdnrs'. Chron.*, **1872**, 1129), and in 1898 S. W. Fitzherbert said it grew like a weed on the island. In the Abbey Gardens he noted *P. tenax*, *P. t. variegatum*, *P. atropurpureum*, *P. colensoi* and *P. guilfoylei*. (*The Garden*, **54**, 473, 1898) Smith, 1964, a gardener, reported *P. tenax* as 'run wild in Tresco', and *P. colensoi* as naturalised on St Martin's. These two, when typical, are distinct enough. Great plants of *P. tenax* with dark-green leaves, some 7 ft long, often with reddish margins, such as those by Pool Road, are very different from *P. colensoi* on St Martin's with narrower, pale-green leaves, about 3–4 ft long, with self-coloured margins. Nevertheless I find it difficult to divide the plants in Scilly between the two 'species' when allowance is made for wide differences in habitat.

P. tenax is an important economic plant, and has been grown as a crop in the Isle of Man, Connemara, and Wigtownshire. Augustus Smith attempted to grow it commercially in Scilly, and Sir Arthur Hill in 1920 urged that a flax industry be set up in co-operation with growers in Cornwall to keep a mill working throughout the year. Nothing came of this.

P. colensoi Hook.f. In 1920 Sir Arthur Hill reported two large plants fully exposed to salt spray at the foot of the western cliffs

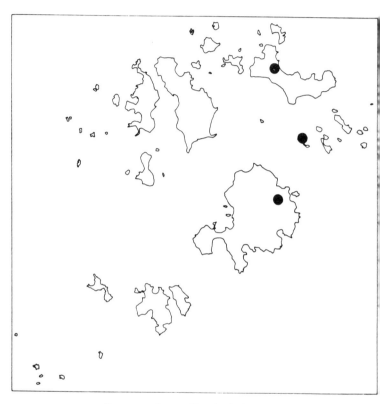

Ruscus aculeatus, Butcher's Broom

on St Martin's, with seedlings springing up in crevices of the granite above and around the original plants. (*Kew Bull.*, **1920,** 170–174, 1920). By 1938 it was well naturalised on cliffs in two bays east of Turfy Hill and in considerable quantity (Lousley, 1939). By 1952 it had spread much further, and is now far west of Turfy Hill on the loose dunes of Great Bay (Plate p 257). In late summer, when much of it is hidden by bracken, it is less impressive, but early in the year the tussocks can be seen to extend for about $\frac{3}{4}$ mile along the cliffs, cliff slopes and dunes, and it is still spreading.

Ruscus L.

R. *aculeatus* L.   Butcher's Broom.   Native.

Laneside, and rough sandy and rocky slopes.   Very rare.
(Map 13, p 262).   M.-.-.-.MN. : -.-.-.-.E.
First record: Bree in Forbes, 1821.
St Mary's: Pelistry Lane. St Martin's: Bree, 1821; St Martin's
Bay, Dallas. Eastern Isles: Great Ganinick, amongst great boulders
near the summit, 1938, Dallas. Seen by me in 1940.

Ornithogalum L.

O. *umbellatum* L.   Common Star of Bethlehem.   Established
alien.
Persisting as weed in bulbfields.   Locally abundant.
M.-.-.-.MN. : -.-.-.-.-.   First record: Lousley, 1967.
St Mary's: bulbfield weed, 1966, P. Z. MacKenzie. St Martin's:
Bulbfields north-west of Pound Lane, 1947, G. Grigson; fields
below Church, 1963; Middle Town in sandy bulbfield, 1959;
plentiful in fields near the Church at Upper Town, 1967; abundant
under bracken in abandoned bulbfields on dune slope by Higher
Town Bay, 1967.

Scilla L.

S. *verna* Huds.   Vernal Squill.   Native.   (Map 14, p 264).
Short turf on cliff slopes.   Local.   M.-.T.B.-. : -.-.-.-.-.
First evidence: 'Scilly Islands, 1813' Herb. Hooker in Herb. Kew
(no doubt collected by W. J. Hooker, who visited Scilly in April
of that year).   First record: Millett, 1852.
St Mary's: Ralfs' *Flora;* Herb. D.-S.; plentiful on cliff from
near Pelistry Bay to Tregear's Porth. Tresco: near Old Grimsby,
Mr Smith, Townsend, 1864; Ralfs' *Flora;* Herb. D.-S. Bryher:
Samson Hill (178) and nearby at intervals; south side of Pool Bay;
Bight Diang, Hell Bay; Shipman Down Head; Gweal Hill, 1947,
G. Grigson.
S. *autumnalis* L. The Misses Millett included this as well as S. *verna*
in their 1852 list, and it is also given in Montgomery, 1854, and repeated
in Ralfs' *Flora*. In the absence of confirmation it seems likely that a late
flowering S. *verna* was mistaken for this species, which would hardly
be out at the time the Milletts' visit ended.

Scilla verna, Vernal Squill

Endymion Dumort.

E. non-scriptus (L.) Garcke   Bluebell.   Cuckoos.   Native.
Under bracken and in short turf on the cliffs; inland on hedge-banks and roadsides.   Very common.
M.A.T.B.MN. : S.AT.H.TN.E.   First record: Millett, 1852
In late spring the sheets of bluebells on the cliff slopes make a colourful display but the distribution is a little patchy. This is probably related to depth of soil, and Grigson has pointed out that on the eastern slopes of Samson bluebells are dominant within the old enclosures (abandoned a century ago) but stop at their boundaries. 'Wild bluebells have always been known locally as Cuckoos' (*Scillonian*, **20**, 69, 1946).

E. hispanicus (Mill.) Chouard . Spanish Bluebell.   Naturalised
alien.
Weed in bulbfields, hedges, roadsides, sandy shores, dunes, in a
copse.
Very common.   M.A.-.B.MN. : -.-.-.-.-.
First record: Lousley, 1967.
St Mary's: Garrison; Parting Carn; Airport; Buzza Hill; Porthcressa; Porth Minick; Bar Point, in *Ammophila;* etc. St Agnes:
common in bulbfields and frequent in hedges. Bryher: occasional,
as under Samson Hill. St Martin's: widely spread from Lower
Town to Brandy Point.
Spanish Bluebell was introduced as a flower crop and has thrived
all too well. It appears to hybridise with *E. non-scriptus,* and
plants believed to be hybrids occur in quantity in an elm copse
above Old Town Church, St Mary's and in fields under Samson
Hill, Bryher.

Muscari Mill.
M. comosum (L.) Mill.   Tassel Hyacinth.   Established alien.
Consolidated dune under Bracken.   Very local.
-.-.-.-.MN. : -.-.-.-.-.   First record: this *Flora.*
St Martin's: St Laurence's Bay, 1967 (872).
This Mediterranean species I have seen thoroughly naturalised
on dunes on the north coast of France, and in the Channel Isles,
and on dunes at Minehead, Somerset. It is likely to persist in
Scilly.

AGAVACEAE
Cordyline Comm. ex Juss.
C. australis Hook. f. While I have no evidence that the 'Palms'
have spread from the places where they are planted, Mr R. C. L.
Howitt tells me that he has often seen seedlings in St Mary's.

JUNCACEAE
Juncus L.
J. compressus Jacq.   Round-fruited Rush.   Error.
This was reported as 'common' by Townsend, 1864 but the specimens
in his herbarium collected from Old Town Marsh in June 1862 are
immature *J. gerardii*. The species is also included by Ralfs, 1879 but he
was probably only following Townsend.

J. gerardii Lois.   Salt-marsh Rush.   Native.
Margins of brackish pools.   Local.
M.A.T.B.-. : -.-.-.-.-.   First record: Townsend, 1864.
St Mary's: Old Town Marsh, June 1862, Townsend, Herb.
Townsend; still there. St Agnes: Pool in Priglis Bay (384). Tresco:
near north corner of Great Pool, (366). Bryher: around the Pool
(335—a form with congested panicles.)

J. bufonius L.   Toad Rush.   Native.
Cart-tracks, ditches, and damp corners of cultivated fields and
gardens.   Abundant.
M.A.T.B.MN. : -.-.-.-.-.   First record: Townsend, 1864.
var. *fascicularis* Koch, which has the flowers in fascicles of 2 or 3,
is recorded from St Mary's by Tellam in Davey, 1909.

J. effusus L.   Soft Rush.   Native.
Wet places.   Local.   M.-.T.-.-. : -.-.-.-.-.
First record: Somerville, 1893 (but see below).
St Mary's: 1862, Townsend in Herb Townsend; Holy Vale;
Porthellick, and Old Town Marsh, Grose; Peninnis cliffs; Lower
Moors (305); Higher Moors (258, 271, 276). Tresco: by Abbey
Pool; Great Pool (358); Middle Down.
A state with the flowers almost sessile, so that they are condensed
into a single dense head ('var.' *compactus* Lej. and Court.) occurs
in Old Town Marsh and by Abbey Pool and has been reported as
the next species.

J. subuliflorus Drej. (*J. conglomeratus* auct.)   Compact Rush.
Native.
Lanesides.   Very rare.
M.-.T.-.-. : -.-.-.-.-.   First record: Lousley, 1967.
St Mary's: High Lane, 1956. Tresco: by Pool Lane, 1956. The
record in Townsend, 1864 is an error as it was based on specimens
of *J. effusus* (see above), and no reliance can be placed on the
citation in Davey, 1909. The species may be more frequent than
the above records indicate.

J. maritimus Lam.   Sea Rush.   Native.
Marshes and margins of pool near the sea, and under cliffs.
Local.   M.-.T.B.-. : -.-.-.-.-.   First record: Townsend, 1864.
St Mary's: Common in the marshes of Old Town and Heugh

Town. Townsend; Lower Marsh, Salmon in Davey, 1909; about Porthellick Pool; rocks below cliff east of Porth Minick. Tresco: Herb. D.-S.; by Abbey Pool. Bryher: a few clumps by the Pool.

var. *atlanticus* J. W. White. (Fig. 7.) This was collected by Townsend from 'Marshes, St Mary's' in June, 1862 (Herb. Townsend) and from St Mary's by A. Somerville in July, 1890 (Hb. Essex and Salmon, *Rep. B.E.C.*, **4**, 164, 1915). J. W. White found it in a 'salt-marsh' on St Mary's, which was Porthellick Marsh, in September 1913, and more mature material was supplied in October. At the present time it is abundant over at least an acre in Old Town Marsh, and also grows in a wet flush on Peninnis Head (spec. 1969, R. Lancaster).

The variety was described by White as differing from the typical species as follows: 'weak and tall, 4 to 5 ft, and the panicle is larger and more diffuse with a lower bract that never exceeds it, and is often not more than a sixth or a quarter its length'. (*J. Bot.*, **52**, 19, 1914; *Rep. B.E.C.*, **3**, 499, 1914; op cit ,**4**, 27, 163-164, 1915.). Except that it is not a rigid plant, it is closely allied to *J. rigidus* Desf. (*Fl. Atlant.*, **1**, 17) which also has very lax panicles up to 20 cm long greatly exceeding the bracts, and is widespread in the brackish marshes of the Mediterranean.

Var. *atlanticus* can be recognised as early as June but the characters are best appreciated from the fully mature panicles in October, or their dead remains the following spring. The weak habit is a function of height rather than due to lack of rigidity in the stems. White noted that it was harvested to provide thatching for a cottage.

J. capitatus Weigel.   Dwarf Rush.   Native.
Very rare.   Only record: St Mary's: Heugh Town, March 27, 1890, C. J. Plumtre in his notebook.
Although this has been seen by no other botanist the record must be accepted. No other rush would be in good condition so early in the year, and I have seen *J. capitatus* in both Cornwall and the Channel Isles in April.

J. acutiflorus Ehrh. ex Hoffm.   Sharp-flowered Rush.   Native·
Marshy meadows and by a pond.   Local.
M.-.-.-.-. : -.-.-.-.   First record: Ralfs, 1879.
St Mary's: Ralfs' *Flora;* Higher Moors (264, 267); Old Town Marsh (533); by pond, Newford.

J. articulatus L.   Jointed Rush.   Native.

Wet places, by pools, marshy meadows, and damp sandy ground. Local.   M.-.T.B.MN. : -.-.-.-.-.   First record: Townsend, 1864

St Mary's: Higher and Lower Moors; Porthloo, Grose. Tresco: south-west corner of Great Pool (356); sandy ground, Herb. Townsend. Bryher: by the Pool. St Martin's: with *Ophioglossum*, Herb. Townsend.

Townsend's specimens are both the dwarf, extensively creeping form characteristic of sandhills.

J. bulbosus L.   Bulbous Rush.   Native.

Tracksides, damp sandy places, and by ponds.   Rather rare.

M.A.T.B.-. : -.-.-.-.-.   First record: Lawson, 1870.

St Mary's: Ralfs' *Flora;* Old Town Marshes; Lower Newford. St Agnes: Herb Townsend (as young state of *J. compressus*). Tresco: Lawson, 1870; Kennelfield, Herb. D.-S.; Lower Pool, Grose   Bryher: Herb. D.-S.

Luzula DC.

L. campestris (L.) DC.   Field Woodrush.   Native.

Heathy places.   Very common.   M.A.T.B.MN. : S.AT.-.-.E.

First record: Ralfs in litt to Townsend, 1876.

L. multiflora (Retz.) Lejeune   Heath Woodrush.   Native.

Boggy places.   Rare.

M.-.T.-.-. : -.-.-.-.-.   First record: Townsend, 1864.

St Mary's: 'Common in the Marshes', Townsend; bogs, Ralfs, 1879. Tresco: near Tresco Abbey, Grose. The records of Ralfs and Grose are for the state with dense subsessile flower-heads (var. *congesta* (DC) Lej.).

AMARYLLIDACEAE

Allium L.

A. babingtonii Borrer   Babington's Leek.   Established alien.

Hedgebanks, field borders, cliffs, dunes, and rough bushy places. (Fig. 6, p, 95).   Common.

M.A.T.B.MN. : -.-.-.-.-.   First record: Lousley, 1940.

St Mary's: Porthellick, Star Castle Hill, and Bar Point, 1939, Grose; hedgebank east of Tolman Point; Old Town; Rose Hill.

St Agnes: field borders near Priglis Bay. Tresco: about 30 plants in hedgerow adjoining School Green 1939, and greatly increased 1956 (368); abundant in field near Dolphin Town; Back Lane and adjoining fields, abundant about Pool Road; shore near New Grimsby; Appletree Banks; Abbey Farm. Bryher: bulbfields near The Town. St Martin's: in several widely scattered fields from Lower Town to Higher Town, 1963.

The dramatic increase since this species was first noticed by Mr Grose and myself in 1939 is quite remarkable and no doubt it now grows in many more stations than those listed above. It appeared first in St Mary's and Tresco in several stations, followed by St Agnes and Bryher, and finally by St Martin's where it was not found until 1963. It evidently has some very efficient dispersal mechanism. This may be by means of the bulbils which are produced in large numbers in the flower heads, but it is difficult to envisage these heavy propagules being carried for considerable distances, and between islands, unless they are transported by birds. It seems more likely that the bulbils' function is the rapid increase of individuals in a colony by short-distance dispersal, but the light wind-born seeds initiate new colonies farther away.

In Cornwall *A. babingtonii* was first known about 1845 from in and near orchards at Grade and Ruan Minor on the Lizard, and was later found in wilder places. It is accepted as native in the west of Ireland, but is unknown outside the British Isles. It is vegetatively active during the winter months and the leaves wither as the stems elongate and flower in June. Like *A. ampeloprasum* its life-history is adapted to a Mediterranean type of climate, but recent monographers of the genus have been inclined to treat it as a variety of *A. scorodoprasum* L. as var. *babingtonii* (Borrer) Rgl.

A. vineale L.   Crow Garlic.   Native.
Sandy places by the sea and roadsides.   Rare.
M.-.T.-.-. : -.-.-.-.-.   First record: Grose ex Lousley, 1940.
St Mary's: Pelistry Bay; Holy Vale and Bar Point, Grose. Tresco: New Grimsby, Grose.
All the plants of this I have seen have the heads composed entirely of bulbils (var. *compactum* (Thuill.) Boreau).

A. roseum L. subsp. *bulbiferum* (DC.) E. F. Warb.   Rosy Garlic.
Weed in bulbfields, roadsides, consolidated dunes, and under bracken.   Local.

M.-.-.-.MN. : -.-.-.-.-.   First record: Vivian in Vigurs, 1913.
St Mary's: 'Abundantly and perfectly naturalised in a field at St Mary's in 1912', Miss Vivian in Vigurs, *J. Roy. Inst. Cornw.*, **19**, 226, 1913. In 1939 I found this attractive species in abundance as a weed on the farms of Tremelethen (228, det. Kew), Salakee and Parting Carn. By 1959 it was common and widespread, and by 1963 abundant almost throughout the island. It is well established on consolidated dune at Bar Point, where it grows with marram-grass and bracken. St Martin's: bulbfields at Higher Town, 1967.
In contrast to other alien Alliums this species has been somewhat slow to spread. It seems that in 55 years it has only recently extended to a second island.

A. triquetrum L.   White Bluebells.   Naturalised alien.
Hedges, roadside gutters, stone walls, field borders and weeds in bulbfields.   Abundant.   M.A.T.B.MN. : -.-.-.-.-.
First record: St Mary's, Perrycoste 1898 in Davey, 1909.
Lady Vyvyan has written a graphic account of the way this plant spreads 'with the speed of an epidemic' in Scilly and of the impossibility of eradicating it (1953, 148-149) and most islanders have abandoned any attempt at control. The way it reached Scilly is not known with certainty but it was probably introduced as a garden plant. Trevellick Moyle has described how in the 90's a Mr Barr (bulb salesman) sold his Uncle Trevellick, who farmed Rocky Hill, bulbs of an *Oxalis* and 'also the lovely smelling one the *Allium*' and remarked 'Better Mr Barr had thrown them in the Thames or Uncle put them on a bonfire' (*Scillonian*, **158**, 101, 1964). The date fits in with Perrycoste's record and by 1939 I found the plant abundant and widespread in St Mary's and Bryher and the south of Tresco, and later on St Agnes and St Martin's. In the Channel Isles it is equally abundant. In Guernsey, for example, it was first noticed in 1847, was not uncommon locally in 1853, and abundant all over the island in 1865. In Cornwall it was first recorded in 1872, spread rapidly, and by 1909 was treated as a native. In fact it is a native of the west Mediterranean and is now well established in the milder parts of England, Wales and Ireland.

A. ursinum L.   Ramsons.   This woodland species has been claimed for Scilly by Davey, 1909 on the evidence of a reference to Borlase which is based on 'Wild Garlick'. The mystery starts with Leland, writing *c.*1533, who says 'diverse of [these] islettes berith wyld garlyk' (Leland, 1907, 191). Borlase had doubts, and wrote 'Wild Garlick grows, as I was informed, in some of the Off-islands, but I met with none'. (Borlase, 1756, 78). Later Woodley said 'Garlick is much cultivated, although it also grows wild'. (Woodley, 1822, 78). The most likely explanation is that the garden Garlic, *Allium sativum* L., so widely grown in the Mediterranean, and to which the soil and climate would be suited, was grown in Scilly as a crop and sometimes got established for a time in waste places. *A. ursinum* is absent from the extreme west of Cornwall and not likely to be found in Scilly.

A. neapolitanum Cyr.   Established alien.
Hedges, wall-tops and bulbfields.   Rare.
M.-.-.MN. : -.-.-.-.   First record: Lousley, 1967.
St Mary's: well established in hedges and wall-tops around Holy Vale, 1959, John Raven; in bracken, Watermill Lane, and elsewhere, 1953. St Martin's: established in bulbfields near the middle of High Town Bay, 1953.
This beautiful plant is less aggressive than *A. triquetrum* and always a relic of cultivation. It has a relatively short flowering period and may be more frequent than the above records suggest.

Nothoscordum Kunth
N. inodorum (Ait.) Nicholson.   Is thoroughly established as a weed in the Abbey Gardens, Tresco. Also in garden of Star Castle, St Mary's, 1969, K. E. Bull.

Ipheion Raf.
I. uniflorum (Grah.) Raf.   Spring Starflower.   Established alien.
Consolidated dune.   Very local.   M.-.-.-. : -.-.-.-.
First record: Moyse in *Scillonian*, **27**, 124, 1952.
St Mary's: Porthloo, in profusion, 1952, Moyse; in 1953 there were scattered patches for about 15 yards on the sea bank, Lousley and B. T. Ward (629); 'hundreds of plants in an area 40 ft x 20 ft', April 1956, J. Matheson; Porthloo, Harry's Walls and Old Town, A. A. Dorrien-Smith (*Scillonian*, **28**, 46, 1953).
I. uniflorum was first reported by Mrs O. Moyse as *Romulea*, and she had first noticed it in 1951. The name was corrected by Major A. A. Dorrien-Smith (*Scillonian*, **27**, 193, 249-250, 1952). It is a native of Argentina and Uruguay which has spread from a

nearby garden. Mrs Moyse said that bulbs were once given to the late Mr John Nicholls of White Cot, who put them in his garden where they increased to the extent of being a pest and spread to the seabanks. Miss E. P. Rogers said it was given to the occupier of 'White Cottage' nearly a 100 years earlier, and suggested that it might have been brought by a sea captain from the River Plate. (*Scillonian*, **28**, 46, 1953). Whatever the earlier history, it is now firmly established at and about Porthloo, and perhaps elsewhere.

Agapanthus L'Hérit.
A. praecox Willd. subsp. *orientalis* (Leighton) Leighton.
Blue Lily.   Established alien.
On consolidated dune.   Local.   (Fig 11, p 251)
M.A.T.-.-. : -.-.-.-.   First record: Lousley, 1940.
St Mary's: established about Old Town. St Agnes: 5 clumps near Gugh Bar, 1963, D. McClintock. Tresco: established on sand-dune near Abbey Farm (367—det Kew as *A. africanus* (L.) Hoffmgg.), 1939—habitat now destroyed; thoroughly established in hollows on Appletree Banks, 1939—still plentiful with blue or white flowers; 'Naturalised and known to visitors as Blue Lily', Pike, A. V., *Gdnrs'. Chron.*, **126**, 208, 1949.
A. orientalis was described by Leighton from Pondoland in 1938 (*J. S. Afr. Bot.*, **5**, 57). It is a larger plant than *A. africanus* with more flowers in the inflorescence and pale- to medium-blue rather than deep blue-violet. In her 1965 revision (The genus *Agapanthus* L'Héritier) she reduced *A. orientalis* to a subspecies of *A. praecox* Willd.

Leucojum L.
L. aestivum L. var. *pulchellum* (Salisb.) Fiori.   Established alien.
M.-.T.-.-. : -.-.-.-.   First record: Lousley, 1967.
St Mary's: in very old bulbfield on the Garrison, 1967, Alf. Page; field near Porthloo Pool, 1970, R. M. Burton. Tresco: established, May, 1929, R. Meinertshagen, Herb. Mus. Brit.; established on Appletree Banks and two other places, 1970, R. M. Burton.

Narcissus L.
Narcissi are to be seen growing out of cultivation in many places and especially round the coasts of the larger islands, by roadsides,

in and on hedges and on waste ground. They persist for a time in fields where cultivation has been abandoned but most of the 'wild' ones have originated from disposal of unwanted bulbs. Probably all the cultivars in present or recent cultivation can be found out of cultivation and the recording of these random throwouts has no scientific value until it can be shown that they persist. Only *N. x. medioluteus* has a history, though others may establish claims in the course of time. For this reason, and to assist those who want to attach scientific names to the flowers they find, the species commonly grown are included here. Those marked with an asterisk were listed by T. A. Dorrien Smith as in Scilly by 1875. Some of these early ones were taken into cultivation from the wild, mainly from bulbs growing round Tresco Abbey, Holy Vale, Rocky Hill, Star Castle and Newford. These were introductions for hortal purposes which found the climate and soil propitious in view of their Mediterranean origins, but only one or two can have been in Scilly before the beginning of the last century (see p 35 and below).

N. tazetta L. Grand Monarque was one of the first to be widely grown, though many so-named are Grand Primo (Hannibal, 1966).

N. ochroleucus Lois. Scilly White. This is said to have grown in Newford orchard 'from time immemorial' (Owen, 1897) but it is unlikely to have been in Scilly before the beginning of the last century.

N. aureus Lois. Soleil d'Or. This very early flowering species is one of the mainstays of the bulb industry but, although it was in cultivation in England before 1600, there is no evidence of an early introduction to Scilly.

N. papyraceus Ker-Gawl. Paper White. An old garden plant cultivated in Italy for centuries but a late introduction to Scilly.

N. x. medioluteus Mill. (*N. biflorus* Curt.) (*N. poeticus x. tazetta*). Primrose Peerless. Cheerfulness. (Plate p 161.) Always sterile, this is believed to have originated as a natural hybrid and it was a common garden plant in England in the sixteenth century. W. J. Hooker collected a specimen labelled 'Scilly Isles' in May 1813 (Herb. Kew) which is before the mass introduction period of Augustus Smith. It may be assumed that he would not have collected such a common garden plant unless he had found it established in a wild habitat and it may have come from Holy Vale, which he is known to have visited, or Tresco. It is now commonly established on the coast and on walls in St Mary's and Tresco and no doubt the other inhabited islands.

N. poeticus L. Pheasant's Eye. In cultivation in England from an early date.

N. bulbocodium L. Tresco: Established, May 1929, R. Meinertzhagen, Herb. Mus. Brit. No doubt a throw-out from the Abbey Gardens.

N. hispanicus Gouan, *N. obvallaris* Salisb. and *N. pseudonarcissus* L. are commonly grown daffodils.

N. gayi (Hénon) Pugsley. Princeps. An important early crop.

N. x. odorus L. (*N. hispanicus x jonquilla*). Campernelli, the Double Daffodil, is said to have been brought to Scilly about 1840 by the captain of a French ship and given to Mrs Gluyas of Old Town, St Mary's.

IRIDACEAE

Sisyrinchium L.

S. californicum (Ker-Gawl.) Ait.f. Tresco: a small colony near the football pitch, Appletree Banks, 1967, found by a Cambridge student and reported by F. H. Perring. This may persist.

S. striatum Smith. Tresco: site of an old garden tip above Appletree Bank, and spreading, 1968, R. Lancaster.

Iris L.

I. foetidissima L. Gladden. Native.
Sand-dunes, rough cliff slopes and lanesides. Uncommon.
M.A.T.-.-. : S.-.-.-.-. First record: Salmon in Davey, 1906.
St Mary's: Old Town, Salmon op cit; near Lifeboat Station, Hugh Town, Dallas; The Garrison; near Innisidgen Point. St Agnes: frequent along shady roadsides and shelter belts. Tresco: Tresco dunes, Smith, 1909; Appletree Banks, 1952; Carn Near; Gimble Porth; under willows by edge of Great Pool. Samson: M. Jaques.

I. pseudacorus L. Yellow Flag. Native.
In marshy meadows, and by a pool. Local.
M.-.T.-.-. : -.-.-.-. First record: Borlase, 1756.
St Mary's: Porthellick, Borlase; common in Higher and Lower Moors. Tresco: about Abbey Pool.

I. xiphium L.   Spanish Iris.   Established alien.
Weed in bulbfields, shore and waste places.   Occasional.
M.-.-.-.MN. : -.-.-.-.-.   First record: Lousley, 1967.
St Mary's: Old Town, on shore, 1959 (833); weed in many bulb-
fields as relic of earlier cultivation. St Martin's: weed in many
bulbfields. Doubtless elsewhere.

Ixia L.
I. paniculata De la Roche   First record: Lousley, 1967.
Tresco: naturalised, 1961, P. F. Yeo, Herb. Kew.

I. speciosa Andr.   Established alien.
Dunes and roadsides.   Frequent.
M.-.-.-.-. : -.-.-.-.-.   First record: Lousley, 1967.
The following records cover a series of garden hybrids which
appear to involve this species.
St Mary's: roadsides near Coastguard Station in several places,
B. M. C. Morgan; laneside near Telegraph (838), 1959; established
in marram grass on shore, in hedgerows, and under bracken in
abandoned bulbfields, Pelistry Bay, (831, 832, 833), 1963; amongst
marram-grass, Bar Point, 1959.

Homeria Vent.
H. breyniana (L.) Lewis. First record: Lousley, 1967. (as *H. collina* Vent.)
Tresco: established along edges of paths in Abbey Gardens where it is
impossible to eradicate, and likely to spread beyond the gardens.

Sparaxis Ker-Gawl.
S. grandiflora (De la Roche) Ker-Gawl.   Established alien.
Roadsides.   Local.
M.-.-.-.-. : -.-.-.-.-.   First record: this *Flora*.
St Mary's: Holy Vale, 1955, Howitt; roadsides near Coastguard
Station in several places, but less frequent than *Ixia*, B. M. C.
Morgan; abundant in lane near Normandy (881), 1967.

Chasmanthe N. E. Br.
C. aethiopica (L.) N.E.Br.   Established alien.
In a swamp.   Rare.
-.-.T.-.-. : -.-.-.-.-.   First record: Lousley, 1967.

Tresco: established in a swamp near Tresco Abbey, and by a track to the Penzance Road, 1963, Lousley, McClintock and Ward. This was in flower at Tresco Abbey on January 6, 1881, *Gdnrs' Chron.*, **1881**, 84.

C. bicolor (Gasp.) N.E.Br.   Established alien.
Laneside and on a wall.   Rare.
M.A.-.-.-. : -.-.-.-.-.   First record: Lousley, 1967.
St Mary's: abundant on field-wall opposite Rose Hill (878), 1967.
St Agnes: lane leading to Wingletang Downs, 1960, J. Russell.

Crocosmia Planch.
C. x. crocosmiflora (Lemoine) N.E.Br.   Montbretia.
Established alien.
Hedgerows, cliff slopes, lanesides.   Common.
M.A.T.-.MN. : -.-.-.-.-.   First record: Lousley, 1967.
St Mary's: thoroughly established and competing with bracken and gorse in many places—e.g. Church Hill; Rocky Hill Lane; Garrison; Holy Vale; Peninnis Lane; hedgerows about Old Town; lane near Pelistry; Carn Leh. St Agnes: hedgerows near Porth Conger and in other parts of the island. Tresco: hedges here and there, and on dune at Appletree Banks and School Green. St Martin's: in a few places in hedgerows.

Gladiolus L.
G. byzantinus Mill.   Whistling Jacks.   Naturalised alien.
Bulbfield weed, hedges, rough ground where it is usually under bracken, consolidated dunes.   Abundant.
M.A.T.B.MN. : -.-.-.-.-.   First record: Lousley, 1940.
This was originally planted by the farmers to supply cut flowers and they now have cause to rue it. The plant persists as a weed in many bulbfields and becomes established and spreads wherever corms have been thrown out. It was well naturalised in several places in 1924 (E. B. Bishop in litt to L. J. Tremayne, 8 June, 1924). *G. byzantinus* has been a garden plant for several centuries, introduced into Europe from Constantinople, but Mr A. P. Hamilton tells me that its wild origin is unknown but probably N. Africa. In Scilly it has several local names. I have heard farmers refer to it as 'Jacks' and 'Rogues' and on St Martin's Geoffrey Grigson was told in 1947 that they called it 'Squeakers'. This no

doubt has the same origin as 'Whistling Jacks' which Mr P. Z. MacKenzie tells me is the usual local name because children use the leaves as reeds to whistle.

In Guernsey *G. byzantinus* has a very similar history. It 'Was grown commercially until superseded by more modern but less hardy gladioli, and it still appears year after year in some meadows, flowering early before the hay is mown'. (Jee, 1967, *Guernsey's Nat. Hist.*, 33.).

ORCHIDACEAE

Spiranthes Rich.

S. spiralis (L.) Chevall.   Autumn Lady's Tresses.   Native.
Sandy turf, but also heathy cliff slopes and wall-tops.   Frequent.
M.A.T.B.MN. : -.-.-.-.-.   First record: Townsend, 1864.
St Mary's: Ralfs' *Flora;* Carn Morval Down; wall between Hugh Town and Old Town; Bar Point, Golf Links, and cricket field on Garrison, O. Moyse; Peninnis, P. Z. MacKenzie. St Agnes: Townsend, op cit; near Troy Town; The Gugh; near Carn Grigland. Tresco: in grass, Herb. D.-S.; slopes near Plumb Hill. Bryher: near Pool in sandy turf; near Puckies Carn. St Martin's: The Plains; cliff near Brandy Point; shore at Lower Town.
The distribution has been given in detail as it illustrates a sharp contrast in habitats. Some, such as Bar Point, Pool and The Plains, are consolidated dune where the soil is likely to be basic. Others, such as Carn Morval Down and Peninnis, are acid heathy ground with calcifuge species. All are near to the sea.

Dactylorhiza Nevski

D. fuchsii (Druce) Soó subsp. *fuchsii.*   Common Spotted Orchid.
Native.
Consolidated dune.   Very rare.
-.-.T.-.-. : -.-.-.-.-.   First record: this *Flora.*
Tresco: a single plant on Appletree Banks, growing with *Calluna vulgaris* and *Pteridium aquilinum*, 1969, M. Walpole.

D. praetermissa (Druce) Soó (*Dactylorchis praetermissa* (Druce) Vermeul.)   Southern Marsh Orchid.   Native.
Marshy meadow and reedbed.   Rare.
M.-.-.-.-. : -.-.-.-.-.   First record: Lousley, 1967.

St Mary's: Lower Moors, 1966, P. Z. MacKenzie who showed me
a large colony in 1967; Higher Moors, 6 plants in thick reed
1967, P. Z. MacKenzie.

This species, like the last, is a recent arrival in Scilly, and no doubt
arose from wind-borne seed blown from Cornwall, where there are
many colonies, or from the Continent.

ARACEAE

Arum L.

A. maculatum L.   Lords and Ladies.

Recorded by Millett, 1852, and by others, but all recent records have
proved to be errors due to confusion with the next species. It is unlikely
that *A. maculatum* has ever occurred in Scilly.

A. italicum Mill. (*A. neglectum* (Towns.)) Ridl.   Late Cuckoo
Pint.   Native.

Hedgebanks, roadsides, hedges, dunes, etc.   Common.

M.A.T.B.MN. : S.-.H.-.-.   First record: Lousley, 1939.

No doubt the earlier records for *A. maculatum*, including Millett,
1852, really refer to this species but it is strange that it was overlooked for so long. Ralfs knew it in Cornwall as long ago as 1877
and it is yet another example of the inadequacy of his work in
Scilly that he overlooked such a conspicuous plant.

Dracunculus Mill.

D. vulgaris Schott.   Dragon Arum.   Established alien.

Very local.   M.-.-.-.-. : -.-.-.-.-.   First record: this *Flora*.

St Mary's: Garrison, 1969, P. Z. MacKenzie; Woolpack Point,
1970 (spec.), K. E. Bull.

Zantedeschia Spreng.

Z. aethiopica (L.) Spreng.   Arum Lily.   Established alien.

Damp places under willows and in scrub.   Local.

M.-.T.-.-. : -.-.-.-.-.   First record: this *Flora*.

St Mary's: naturalised in scrub near the Lifeboat Station, Howitt;
Pelistry Bay, naturalised, C. T. Prime. Tresco: abundant beneath
willows along the north side of Great Pool, Howitt; spreading in
Well Top, by Abbey Gardens, 1968, R. Lancaster.

LEMNACEAE
Lemna L.

L. minor L.   Common Duckweed.   Native.

Floating in ponds and ditches.   Rather rare.

M.A.T.B.-. : -.-.-.-.-.   First record: Townsend, 1864.

St Mary's: Holy Vale, Grose; ditch near Rose Hill; pool in Water-mill Valley; Old Town Marsh; Porthellick; Lower Newford; Normandy. St Agnes: Wingletang Downs; small pool near the Nag's Head. Tresco: pool near Abbey. Bryher: in one small pond at Pool.

BROMELIACEAE
Fascicularia Mez.

F. pitcairniifolia (Verlot) Mez.   Established alien.

Consolidated dune, roadsides and quarry.   Local.

M.-.T.-.-. : -.-.-.-.-.   First record: this *Flora*.

St Mary's: Bank of quarry near Bant's Carn Farm, 1962, Howitt; roadside, Old Town, 1967, L. J. Wood. Tresco: Appletree Banks. 'Plants sent from Kew to Tresco are stated by Mr Dorrien-Smith to flower there annually in the open air', 1906 (*Bot. Mag.*, *t*.8087 as *Rhodostachys pitcairniifolia* Benth.). Many of the plants on Appletree Banks grew from cuttings put there by Major A. A. Dorrien-Smith but it may be spreading, and Dr H. Heine tells me that on an island off the French coast with a similar climate it is thoroughly naturalised.

SPARGANIACEAE
Sparganium L.

S. erectum L.   Branched Bur-Reed.   Native.

Marshes.   Very rare.   M.-.-.-. : -.-.-.-.-.

First record: Ralfs in litt to Townsend, 1876.

St Mary's: Marsh at Old Town, Ralfs, loc cit, and 1879; 'Bog, St Mary's', Marquand, 1893; Lower Moors, 1966, P. Z. MacKenzie.

TYPHACEAE
Typha L.

T. latifolia L.   Greater Reedmace.   Native.

Margins of large ponds.   Rare.

M.-.T.-.-. : -.-.-.-.-.   First record: Lousley, 1967.

St Mary's: Porthellick Pool, about 12 'maces', 1967, F. Russell Gomm. Tresco: near Abbey, 1939, Grose; Great Pool by Pool Road, 1939; Duckery, 1952.

CYPERACEAE

Eriophorum L.

E. angustifolium Honck.　　　Common Cottongrass.　　　Native.
Bogs.　　　Rare.

M.-.-.-.-.　:　-.-.-.-.-.　　　First record: Millett, 1852.
St Mary's: abundant in Old Town marshes, Townsend; Higher Moors, Herb. D.-S.; bog near Tremelethen (273; 415) steadily decreasing since 1939, and now very scarce; Porthellick, 1955, Howitt.

Two very different forms of this species have occurred in the little bog on Higher Moors. One has bristles some 35 mm in length (415) and is the usual form in Britain generally. The other, which was less plentiful, had bristles only 5–7 mm long in mainly subsessile spikelets (273) and superficially resembled *E. latifolium*. For a fuller discussion of this plant see Lousley 1940, 158.

E. latifolium Hoppe. This species was recorded by Lawson, 1870 who no doubt mistook the form of *E. angustifolium* mentioned above for this species. *E. latifolium* is unlikely to be found in Scilly as there are no suitable base-rich marshy habitats.

The subsequent history of Lawson's record is an example of the confusion which has dogged earlier work. All Lawson's records were entered by Townsend in a copy of his own paper which is now at the South London Botanical Institute. This manuscript he lent to Ralfs who copied the record for *E. latifolium* into his *Flora* as on Townsend's authority. Davey, 1909 copied it from Ralfs' *Flora* with no mention of Lawson. So years later Townsend, a leading botanist of his day, is credited with a record he had never made and which is incorrect.

Scirpus L.

S. maritimus L.　　　Sea Clubrush.　　　Native.
Slightly brackish marshes and pools.　　　Locally abundant.
M.A.T.-.-.　:　-.-.-.-.-.　　　First record: Townsend, 1864.
St Mary's: common in St Mary's marshes, Townsend; Old Town, Herb. Downes; Old Town Marshes; Porthellick. St Agnes: about the pool in Priglis Bay. Tresco: south end of Great Pool.

At Porthellick J. D. Grose noted plants with simple congested panicles of sessile spikes (var. *conglobatus* S. F. Gray).

S. tabernaemontani C. C. Gmel.   Glaucous Bullrush.   Native?
Recorded by Somerville, 1893, but not noted by any other botanist.
Likely habitats include Porthellick Pool, St Mary's and Great
Pool, Tresco.

S. setaceus L.   Bristle Clubrush.   Native.
Wet, sandy places.   Very rare.
M.-.T.-.-. : -.-.-.-.-.   First record: Townsend, 1864.
St Mary's: fresh water trickle near Block House Point, 1948,
Ribbons and Wanstall; Porthellick, 1955; Old Town Marsh, Howitt;
wet flushes between Pelistry Bay and Bar Point, 1969, Lancaster.
Tresco: damp peaty depressions on Tregarthen Hill, 1969, Lancaster.
Townsend, 1864 gave the species as common but this was an
afterthought as his 1862 manuscript has 'Not unfrequent?' Ralfs'
Flora also says 'common' but this was probably merely copying
Townsend.

S. cernuus Vahl.   Slender Clubrush.   Native.
Wet bare places on mud or sand.   Frequent.
M.A.T.-.-. : -.-.-.-.-.   First record: Townsend, 1864.
St Mary's: Ralfs' *Flora;* Lower Moors near Rosehill (304); Old
Town Marsh, Polunin; Higher Moors (277). St Agnes: 1923,
Herb. Downes; Wingletang Downs (436); by pool, Priglis Bay.
Tresco: Abbey Pool (509, 726).

S. fluitans L.   Floating Clubrush.   Native.
Probably floating in ditches.   Very rare.
M.-.-.-.-. : -.-.-.-.-.   First record: Townsend, 1864.
St Mary's: Ralfs' *Flora;* June 1863 (should be 1862), Herb. Townsend.
Townsend listed this as 'Frequent in the marshes', which cannot
have been true as there is no evidence that any other botanist has
seen it.

Eleocharis R.Br.
E. quinqueflora (F. X. Hartmann) Schwarz. Error. Townsend, 1864 gives
this as 'common' but a specimen in his herbarium which was first so
labelled has been corrected to *E. multicaulis* and his 1862 manuscript
altered accordingly. Ralfs' *Flora* gives St Mary's and Tresco for the
species, but this is no doubt based on material collected by Curnow
from these islands on May 26, 1876 which was first named '*Scirpus
pauciflorus*' and later determined by A. Bennett as *E. uniglumis*.

E. multicaulis (Sm.) Sm. Many-stalked Spikerush. Native.
Peaty places in the marshes and by ponds. Rare.
M.-.T.-.-. : -.-.-.-. First record: Townsend, 1864.
St Mary's: June, 1862, Townsend in Herb. Townsend; Higher
Marsh, 1905, Salmon in Davey, 1909; Old Town Marsh, 1922,
Herb. Downes; Old Town Marsh (655); Porthellick, 1955, Howitt.
Tresco: Abbey Pool, 1938.

E. palustris (L.) Roem. and Schult. Marsh Spikerush. Native.
Marshes, ditches and pondsides. Frequent.
M.-.T.B.-. : -.-.-.-. First record: Townsend, 1864.
St Mary's: Ditch by roadside, 1862, Townsend in Herb. Townsend;
Old Town Marsh and Porthloo, Grose; pool on Buzza Hill, Dallas;
Old Town Marsh (*180); pool at top of Watermill Lane (*288).
Tresco: Ralfs' *Flora;* Great Pool, Grose, north corner of Great
Pool (365). Bryher: near *Pool, 1952, A. Conolly.

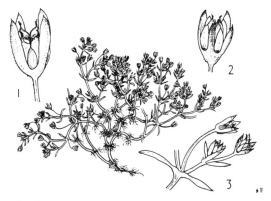

Fig 12 *Crassula decumbens*, a small bulbfield weed introduced from
   South Africa. Whole plant x1; fruit and flower x5; group of
   leaves, flower and fruit x2½

The three gatherings marked with an asterisk have been determined
by S. M. Walters as subsp. *vulgaris* Walters (subsp. *palustris* of
the *Plant List*), and this is likely to be the only subspecies occurring
in Scilly.

E. uniglumis (Link) Schult.   Slender Spikerush.   Native.
By ponds.   Rare.   M.A.T.-.-. : -.-.-.-.-.
First record: W. Curnow in Davey, 1903, and Davey, 1909.
St Mary's: Higher Moor, teste Bennett, op cit. St Agnes: by the pool, Priglis Bay. Tresco: Abbey Pool (183, teste Walters); near extreme north corner of Great Pool (365, teste Walters). This is not the typical *E. uniglumis* of Eastern England but '*uniglumis* of sorts' and a difficult plant.

Rhynchospora alba (L.) Vahl, White Beak-sedge, was recorded by P. Z. MacKenzie in 1966 but no specimen was collected and confirmation is required (Lousley, 1967).

Carex L.
C. laevigata Sm.   Smooth Sedge.   Native.
Rushy marshy meadows.   Rare.
M.-.-.-.-. : -.-.-.-.-.   First record: Townsend, 1864.
St Mary's: Near Old Town Marsh, Townsend; Higher Bog, 1876, W. Curnow, Herb. Curnow; Higher Marsh, 1923, Downes in Herb. Downes; Higher Moors (268; 663).
In Scilly, as in Cornwall, *C. laevigata* grows in open marshes in contrast to the more usual woodland habitats elsewhere. Townsend recorded his plants as 'var. *schraderi* Woods', which is said to differ from the typical species in having the glumes awned. The material in his herbarium, collected from marshy ground, St Mary's, June 1862 has 'Beak of fruit nearly or quite smooth. Glumes with long setaceous point' and the plants were 6 ft tall. My own specimens are similar but the characters do not seem significant.

C. binervis Sm.   Moor Sedge.   Native.
Heathy rather acid places—especially by tracks.   Local.
M.-.-.-.-. : -.-.-.-.-.   First record: Townsend, 1864.
St Mary's: Ralfs' *Flora;* Golf Ground, 1923, Downes in Herb. Downes; Bar Point, Grose; McFarland's Down (464); near Innisidgen; Porth Hellick, Grose; Aerodrome, Polunin.

C. lepidocarpa Tausch subsp. *lepidocarpa*. Error. This was recorded by Tellam in *Rep. Bot., Record Club*, **1873–7**, 244, 1878, but his specimen collected on St Mary's, June 29, 1877, is *C. demissa* (Herb. Mus. Brit.). Townsend added *C. lepidocarpa* to an annotated copy of his paper but the specimen in his herbarium first labelled '*flava minor* mihi', and altered to *C. lepidocarpa*, is *C. demissa*.

C. demissa Hornem.   Common Yellow Sedge.   Native.
Marshy meadows.   Local.
M.-.-.-.-. : -.-.-.-.-.   First record: Townsend, 1864 (as *C. oederi*)
St Mary's: Higher Moor, 1923, Downes in Herb. Downes; Higher
Moor, 1876, W. Curnow, Herb. Curnow; Old Town Marsh, 1922
H. Downes in Herb. Mus. Brit.; Old Town Marsh (652); Porthellick
and Old Town, 1955, Howitt.
In spite of all the previous taxonomic confusion, and the complica-
tion of name changes, it seems clear that this is the only sedge of
the '*flava*' group which has been found in Scilly. In addition to
the specimen already cited, there is material labelled 'Scilly, June
1862' in Herb. Townsend named '*minor* Townsend' (which is this
species) in Townsend's own handwriting.
C. flava L. In Townsend, 1864 as 'Common in the marshes'. Error—
C. demissa intended.

C. serotina Merat. Error. This, as *C. oederi* Ehrh, was also given in
Townsend, 1864 as 'common in the marshes'. In his manuscript he
added 'Requires confirmation' but Ralfs and Davey followed the
uncorrected printed record.

C. sylvatica Huds.   Wood Sedge.   Native.
In woodland.   Very rare.
-.-.T.-.-. : -.-.-.-.-.   First record: Lousley, 1939.
Tresco: south part of Tresco Gardens; still there, 1955, Howitt.
This is on, or very near, the site of Abbey Wood, the only woodland
in Scilly in recent history.

C. riparia Curt.   Great Pond Sedge.   Native.
Marsh.   Very rare.
M.-.-.-.-. : -.-.-.-.-.   First record: Townsend, 1864.
St Mary's: Old Town Marsh, Townsend (recorded with doubt);
Porthellick, 1955, Howitt. Also recorded without locality as
found by Miss C. Vivian in 1912, Thurston and Vigurs, 1922.

C. pendula Huds.   Pendulous Sedge.   Established alien?
In a garden and on a roadside.   Rare.
M.-.T.-.-. : -.-.-.-.-.   First record: Lousley, 1967.
St Mary's: opposite Rose Hill, 1967. Tresco: plentiful below the
Eastern Rockery, Abbey Gardens.

C. panicea L.   Carnation Sedge.   Native.
In marsh, probably where acid.   Very rare.
M.-.-.-.-. : -.-.-.-.-.   First record: Lousley, 1940.
St Mary's: Old Town Marsh, June 1923, Downes in Herb. Downes
(as *C. flacca*).

C. flacca Schreb.   Glaucous Sedge.   Native.
Consolidated dune.   Rare.
-.-.T.-.-. : -.-.-.-.-.   First record: Lousley, 1967.
Tresco: on maritime turf, May 1929, R. Meinertzhagen, Herb.
Mus. Brit.

C. pilulifera L.   Pill Sedge.   Native.
'Marshes'.   Rare.
M.-.-.-.-. : -.-.-.-.-.   First record: Ralfs, 1879.
St Mary's: Higher Marsh, Ralfs' *Flora* (see also Marquand,
1893 and Davey, 1909). A rather unsatisfactory record. This
species is common in Cornwall, and is likely to be found by tracks
on the heaths, but the only record is from Higher Marsh.

C. nigra (L.) Reichard.   Common Sedge.   Native.
Marshy meadows.   Locally plentiful.
M.-.T.-.MN. : -.-.-.-.-.   First record: Townsend, 1864.
St Mary's: Marshes, Townsend, 1864; common, Ralfs' *Flora;*
Old Town Marshes, locally abundant (397): Lower Moors near
Porth Mellon. Tresco: Townsend. St Martin's: Townsend.

C. paniculata L.   Tussock Sedge.   Native.
Wetter parts of the marshes.   Locally abundant.
M.-.-.-.-. : -.-.-.-.-.   First record: Smith, 1907.
St Mary's: Old Town Marsh, 1922 and Higher Marsh, 1923,
Downes in Herb. Downes and Thurston and Vigurs, 1922; Higher
Moors, Smith, 1909; Higher Moors, in quantity (275; 272).
W. W. Smith (later Sir William Wright Smith) in August 1906
described the exceptionally fine tussock development of this
species at Higher Moors, where the tussocks were associated with
Phragmites and *Pteridium* to form growths recalling features of a
tree-fern. He gave measurements of two individuals as: No. 1.
Height of stump 5½ ft, total height 10 ft, circumference of stump
at thickest 8⅔ ft. No. 2. Height of stump 4ft, girth at 3 ft–7 ft,

circumference of crown 12 ft. The parts of the *Pteridium, Phragmites* and *Carex* were evenly distributed throughout the tussocks. These tussocks carried other epiphytic species (Smith, W. W., 1907: Note on a Peculiar Tussock Formation, *Trans. Bot. Soc. Edinb.*, **23**, 234–235, tab. 1.). Attempts at drainage, and destruction by toxic chemicals and fire, have reduced the size of the tussocks, but the formation, which can be seen on Higher Moors near Tremelethen, is still of great interest.

Two of Downes' herbarium sheets have been labelled var. *brevis* Asch. and Graebn. and var. *simplicor* And., but they are merely immature specimens of *C. paniculata*.

C. otrubae Podp.   Fox Sedge.   Native.
Marshes.   Rare.
M.-.-.-.-.   : -.-.-.-.-.   First record: Lousley, 1940.
St Mary's: rushy meadow at Higher Moors (278); margin of large pool at Tremelethen, 1958, R. P. Bowman (spec.); Lower Moors, east side, 1968, P. Z. MacKenzie (spec.).

C. arenaria L.   Sand Sedge.   Native.
Sandy coasts and dunes, tops of stone-walls, rough ground on cliff slopes and also inland on hedges.   Abundant.
M.A.T.B.MN. : S.AT.H.TN.E.   First record: Townsend, 1864.
First evidence: St Mary's, 1845, Hambrough in Herb. S.L.B.I.
There is probably nowhere in the British Isles where *C. arenaria* may be seen in greater quantity and under a wider range of conditions than in Scilly. It is by no means confined to the coast, and also grows about as far from the sea as it is possible to get, as on field walls at Normandy, and on high ground, as on the top of South Hill, Samson.

C. ligerica J. Gay, an allied species, is accepted for Scilly by Ascherson and Graebner (*Syn. Mitteleur. Fl.*, **2/1**, 32, 1902) and by the great monographer of the genus, Kükenthal (*Pflanzenreich*, **4-20**, *Cyperaceae-Caricoideae*, 139, 1909). This originated with specimens collected by J. Cunnack on St Mary's in July 1878 which Arthur Bennett, after consulting Boeckler and Babington, published as *C. ligerica* Gay, new to the British flora (*J. Bot.*, **22**, 27–28, 1884). Further material was collected by Cunnack from the 'Slope of Star Castle' on June 16, 1883 and by A. Ley from St Mary's (probably the same locality) on July 3, 1883 (Herb. Mus. Brit., Herb. Kew, Herb. S.L.B.I., etc.; *Rep. B.E.C.*, **1**,

192, 188), and Bennett exhibited material at a meeting of the Linnean Society on December 6, 1883. Cunnack's specimens are of a slender drawn-up state of *C. arenaria*, and similar material still grows in the tangle of bramble and bracken on the slopes below Star Castle. G. C. Druce referred Cunnack's plants to *C. arenaria* var. *remota* Marss. but they are nothing more than a habitat state (*Rep. B.E.C.*, **5**, 793, 1920). Another form has the spike reduced to one or two spikelets (192b) and occurs in a large uniform colony amidst masses of the typical species (192a) on Annet.

C. divulsa Stokes.   Native.
Under trees.
-.-.T.-.-. : -.-.-.-.-.   First record: Lousley, 1939.
Tresco: wooded lower part of Abbey Gardens (191; 351); abundant by lane near Dial Rocks (783).

C. spicata Huds.   Spiked Sedge.   Doubtful.
C. muricata L. was recorded by Somerville, 1893 on Arthur Bennett's determination, but this was before the plant now known under this name (*C. pairaei* F. W. Schultz) was distinguished in Britain. A very young specimen from Old Town Marsh, 1923 in Herb. Downes may well be *C. spicata* but is too immature for definite determination. It is certain that a sedge of this group has occurred but it is not known which species.

C. echinata Murr.   Star Sedge.   Native.
Acid parts of the marshes.   Rare.
M.-.-.-.-. : -.-.-.-.-.   First record: Townsend, 1864.
St Mary's: Common in the marshes, Townsend; Higher Bog, 1876, W. Curnow in Herb. Curnow; Higher Marsh, Downes in Herb. Downes; bog on Higher Moors (662; 799).

GRAMINEAE

Phragmites Adans.
P. australis (Cav.) Trin. ex Steud. (*P. communis* Trin.).   Reed.
Native.
Marshes and around pools.   Locally abundant.
M.-.T.-.-. : -.-.-.-.-.   First record: Townsend, 1864.
St Mary's: Higher Moors; Lower Moors—Rose Hill, Porthloo, etc. Tresco: abundant round Great Pool and forming large reedbeds.

Cortaderia Stapf
C. selloana (J. A. and J. H. Schult.) Aschers. and Graebn.   Pampas-grass.   Established alien.
Consolidated dune and roadside.   Local.

-.-.T.-.-. : -.-.-.-.-. First record: Lousley, 1967.
Tresco: Appletree Banks, 1959; by Pool Road, 1956. Many of these plants have no doubt been planted, but persist without attention, and others are in places where they seem likely to have grown from seed.

Molinia Schrank
M. caerulea (L.) Moench. Purple Moor-grass. Native.
Wet heathy places. Local.
M.-.T.-.-. : -.-.-.-.-. First record: Lawson, 1870.
St Mary's: Ralfs' *Flora;* by path from Innisidgen Carn to Bar Point (280); about Porthellick. Tresco: Townsend, MS.; Ralfs' *Flora;* Lawson, 1870; plentiful by Abbey Pool.

Sieglingia Bernh.
S. decumbens (L.) Bernh. Heath Grass. Native.
Peaty places in the marshes, heathy ground with *Calluna*. Local.
M.A.-.B.MN. : S.AT.-.-.-. First record: Townsend, 1864.
St Mary's: Old Town Marshes (654); Higher Moors; Peninnis, Airport, Porthellick, Innisidgen, Toll's Island, etc. St Agnes: The Gugh; Wingletang Downs; by Priglis Pool (690). Bryher: Shipman Head. St Martin's: St Martin's Bay; Chapel Down, Gomm. Samson: Grose. Annet: bare ground between *Calluna* bushes, 1948, Ribbons and Wanstall.

Glyceria R.Br.
G. fluitans (L.) R.Br. Flote-grass. Native.
By ponds and ditches. Local.
M.-.-.-.-. : -.-.-.-.-. First record: Townsend, 1864.
St Mary's: Hugh Town Marshes, etc., Townsend; Porthloo (396, teste C. E. Hubbard); Porthloo (800, teste A. Melderis); marsh, Tremelethen, Polunin; Porthloo, 1952, Grose as 'forma (or var.) *triticea* (Fries)' det. C. E. Hubbard.

G. plicata Fr. Sweet-grass. Native.
In a marsh. Very rare.
M.-.-.-.-. : -.-.-.-.-. Only record: Townsend, 1864.
St Mary's: St Mary's Marsh, Townsend. In his paper Townsend, who made a very careful study of this genus, distinguishes *G. plicata* from the other two species and sets out the characters by which he did so. The record must therefore be accepted.

(*left*) Bear's Breech, *Acanthus mollis*, growing on St Agnes in a colony which dates from c 1800; (*right*) Shore Dock, *Rumex rupestris*, a rare plant common on the coast of Scilly.

G. declinata Bréb.   Small Flote-grass.   Native.

Ponds, ditches and other wet places in the marshes.   Local.

M.-.-.-.-. : -.-.-.-.-.   First record: Townsend, 1864.

St Mary's: 'Ditches and wet places in the marshes', Townsend, 1864; St Mary's Marshes, June 1862, Herb. Townsend and Herb. Babington; Higher Moors near Tremelethen (798, confirmed A. Melderis); Newford Ponds.

It was Townsend's visit to Scilly which first brought this species to the notice of botanists as a British plant. In his 1864 paper he described it as *G. plicata* var. *nana*. Later he identified this with *G. declinata* described by Brébisson from Normandy, under which name it is now known. (Townsend, F., *Fl. Hants.*, ed. 1, 508–509, 1883; ed. 2, 646–647, 1904). The last mentioned account describes the circumstances under which he first found the plant in Scilly: 'Small forms of *G. fluitans* occurred in plenty, growing with *G. declinata*, but *G. plicata* was absent'. It is therefore evident that his *G. plicata* was from a different locality on St Mary's.

x *G. fluitans*. Specimens collected by me from Porthloo Pool, St Mary's on September 15, 1936, with imperfect pollen, are a hybrid of *G. declinata* for which C. E. Hubbard suggests *G. fluitans* as the other parent.

Festuca L.

F. pratensis Huds.   Meadow Fescue.   Native?

Edges of fields.   Rare.

M.-.T.-.-. : -.-.-.-.-.   First record: Lousley, 1967.

St Mary's: Edge of bulbfield at Green. Tresco: near the Abbey, 1939, Grose.

This species has not yet been found in a habitat where it can be safely claimed as native, and it may be introduced with seed for temporary leys.

F. rubra L.   Red Fescue.   Native.

Sandy pastures and dunes, sandy cliffs, roadsides, and edges of bulbfields, walltops.   Abundant.

M.A.T.B.MN. : S.AT.H.TN.E.   First record: Townsend, 1864.

The common plant is subsp. *rubra*, to which the above records refer. My specimens from Priglis Bay, St Agnes (200), and Great Innisvouls (322) were determined by W. O. Howarth as *F. rubra* L. var. *glaucescens* (Hegets. and Heer) Richt. and this is probably common round the coast.

The alien subsp. *commutata* Gaudin was abundant in 1943 in bulbfields near Parting Carn, St Mary's (843) and was probably introduced with a seed mixture.

F. ovina L.   Sheep's Fescue.   Native.
Dry pastures, heaths, cliffs, etc.   Common.
M.A.T.B.MN. : -.-.H.TN.E.   First record: Townsend, 1864.

Lolium L.
L. perenne L.   Perennial Rye-grass.   Native.
Meadows, tracksides, roadsides and gateways.   Common.
M.A.T.B.MN. : -.-.-.TN.-.   First record: Townsend, 1864.

L. multiflorum Lam.   Italian Rye-grass.   Casual.
Clover leys, bulbfields, field borders, etc.   Rare.
M.-.T.-.MN. : -.-.-.-.   First record: Townsend, 1864.
St Mary's: In a clover field near Hugh Town, Townsend; field border near Holy Vale (223—det. C. E. Hubbard); Gateway to bulbfield near Tremelethen; Four Lanes' End, 1970. Tresco: weed in Abbey Gardens, Lancaster. St Martin's: plentiful in bulbfields at Higher Town and elsewhere.
A century ago Italian Rye-grass was a constituent of the clover mixture sown on barley-land (Scott and Rivington, 1870), whereas now it is sown on bulbfields in years when they are rested from Narcissi.

Vulpia C. C. Gmel.
V. membranacea (L.) Dumort. Recorded by Somerville, 1893. The record is supported by two specimens labelled 'Scilly, v.–c.1, July '90' (Herb. Edinb.) but there is no further information.

V. bromoides (L.) Gray.   Squirrel-tail Fescue.   Native.
Dry, and usually sandy, places and also a bulbfield weed.
Very common.
M.A.T.B.MN. : S.-.H.-.-.   First record: Townsend, 1864.
This is a native species which has probably increased considerably since the introduction of the bulb industry. As a weed in the bulbfields it is larger and often abundant.

V. myuros (L.) C. C. Gmel.   Rats-tail Fescue.   Native.
Dry sandy ground, and common as a bulbfield weed.   Common.
M.A.T.-.MN. : -.-.-.-.-.   First record: Lousley, 1939.
In Scilly as elsewhere this species is seen mainly in man-made
habitats. In bulbfields it is now common and luxuriant, having
increased considerably during the last 20 years, and it also occurs
on roadsides and waste places.

Puccinellia Parl.
P. maritima (Huds.) Parl.   Sea Meadow-grass.   Native.
Brackish marsh.   Extinct.
M.-.-.-.-. : -.-.-.-.-.   First record: Curnow in Ralfs, 1879.
St Mary's: 'Salt portion of Higher Bog, May 28, 1877', W. Curnow
in 'Scrap-book' at Tresco Abbey; Higher Marsh, W. Curnow in
Ralfs' *Flora*.
This was found at Porthellick and the habitat having become less
salt is probably no longer suitable.

Catapodium Link.
C. rigidum (L.) C. E. Hubbard.   Hard Fescue.   Native.
Dry sandy places and walls.   Rare.
M.A.T.-.MN. : -.-.-.-.-.   First record: Townsend, 1864.
subsp. *rigidum*—St Mary's: base of wall near Hugh Town Church
(195). St Agnes: Ralfs' *Flora*. St Martin's: 1956, Howitt.
subsp. *majus* (C. Presl) Perring and Sell—St Mary's: cultivated
field, 1934, Miss E. S. Todd, det. C. E. Hubbard, *Rep. B.E.C.*, **11**,
290, 1937; Star Castle Hill and near Old Town, Grose; bulbfield
near Hugh Town. Tresco: Abbey Grounds (194).
This 'subspecies' is a taller plant, with wider leaves, and spreading
panicles. It remains constant in garden soil and has an interesting
distribution in south-west England, Ireland, and the Channel
Isles, being known abroad in south-west Europe and much of the
Mediterranean. In my 1939 paper it is discussed as *Scleropoa
rigida* (L.) Griseb. var. *major* (J. B. Presl) Lousley, but further
experience in the field suggests that it only occurs in places where one
would expect a grass to be more luxuriant and that its characters may
be related to climatic conditions. Experimental work is much needed.

C. marinum (L.) C. E. Hubbard.   Darnel Fescue.   Native.
Dry places near the coast.   Frequent.
M.A.T.B.MN. : S.AT.H.TN.-.   First record: Millett, 1852.

Poa L.
P. annua L.   Annual Meadow-grass.   Native.
Almost ubiquitous but especially on cultivated land.   Very common.
M.A.T.B.MN. : S.AT.-.TN.-.   First record: Townsend, 1864.
An interesting variation, which C. E. Hubbard suggests is probably forma *aquatica* (Aschers.) Junge (var. *aquatica* Aschers.) occurs on St Mary's in Old Town Marsh (398).

P. infirma Kunth.   Early Meadow-grass.   Native.
Open sandy ground on the coast and especially on sandy tracks. Also on granite carns associated with *Plantago coronopus*.
Common.   (Fig 7, p 125.)
M.A.T.B.MN. : -.-.-.-.-.   First record: Raven, J. E., *Watsonia*, **1,** 356, 1950.
St Mary's: common. St Agnes: apparently rare; cliffs near St Warna's Cove (610), D. McClintock. Tresco: common except in north of island. Bryher: local; hillocks at Pool (560); Rushy Bay. St Martin's: common.
This grass flowers in March and early April and in most seasons is difficult to find after the middle of that month. Its late discovery, and the lack of records from the uninhabited islands, are explained by the rarity of visits from botanists so early in the year. *Poa infirma* is abundant in the Channel Islands, where it has been known since 1910 (and was collected much earlier), but was not detected in Scilly or on the mainland of Cornwall until Mr John Raven found it in April 1950.

P. pratensis L.   Smooth Meadow-grass.   Native.
Dunes, pastures, wall tops.   Frequent.
M.A.T.B.MN. : -.-.-.TN.E.   First record: Tellam in Davey, 1909.
P. angustifolia L. Doubtful. In 1957 I made a note in my field records that I saw this on a cottage wall near the Church, Hugh Town, St Mary's, but I did not keep a specimen. In 1966 Mrs B. Lucas listed this as common, which it certainly is not, but I have not seen a specimen.

P. subcaerulea Sm.   Spreading Meadow-grass.   Native.
Sandy places and marshes.   Probably common.
M.-.-.B.MN. : -.-.H.-.E.   First record: Townsend, 1864.

St Mary's: Townsend, 1864; Porthellick; Tremelethen, R. P. Bowman. Bryher: Pool Bay. St Martin's: Townsend, 1864; Lower Town. St Helen's: near Pest House (816, conf. Melderis). Eastern Isles: Middle Arthur.

There is no doubt that this species is heavily under-recorded. Much of my field work was done before this was properly separated from *P. pratensis* and few observers have distinguished the two species.

P. trivialis L.　　Rough Meadow-grass.　　Native.
Marshes, meadows, bulbfields, and roadsides.　　Abundant.
M.A.T.B.MN. : -.-.-.-.-.　　First record: Townsend, 1864.
In Scilly this flowers in June and July, when *P. pratensis* and *P. subcaerulea* are mainly over. Luxuriant forms with large panicles of large spikelets occur in Old Town Marsh and as weeds in bulbfields. J. D. Grose recorded var. *laevis* Lej. and Court. (var. *glabra* Döll) from Star Castle Hill, St. Mary's.

Dactylis L.
D. glomerata L.　　Cocksfoot.　　Native.
Meadows, roadsides, borders of bulbfields and cliffs.　　Common.
M.A.T.B.MN. : S.AT.H.TN.E.　　First record: Townsend, 1864.

Cynosurus L.
C. cristatus L.　　Crested Dogstail Grass.　　Native.
Pastures and roadsides.　　Frequent.
M.A.T.B.MN. : -.-.-.TN.-.　　First record: Townsend, 1864.
Usually in turf, and most plentiful in damp places about pools and marshes. The wiry stems were used in the straw-plait industry, 1800–1880 (Hayward, L. A., *Scillonian*, **13**, 44–46, 1939).

C. echinatus L.　　Rough Dogstail Grass.　　Naturalised alien.
Dry open slopes and bulbfields.　　Local.
M.-.T.-.-. : -.-.-.-.-.　　First record: Lousley, 1939.
St Mary's: Buzza Hill, 1954 (647), increasing 1963 and 1967; Sea Ways, 1962 (827), Howitt. Tresco: Farmbank, above High Water mark, 1933, Herb. D.-S.; abundant in bulbfield near Back Lane, 1940.

Briza L.

B. *minor* L.   Small Quaking-grass.   Colonist.
Bulbfield weed.   Abundant.
M.A.T.B.MN. : -.-.-.-.-.   First record: Lawson, 1870.
The dainty grass is a native of the Mediterranean region and is probably an ancient introduction. It is an annual, which thrives under conditions of bulbfield cultivation and occurs in much greater quantity in Scilly than I have seen it elsewhere.

B. *maxima* L.   Great Quaking-grass.   Established alien.
Roadsides, walltops and arable land.   Rare.
M.-.T.-.MN. : -.-.-.-.-.   First record: Downes in Thurston and Vigurs, 1922.
St Mary's: roadside, Downes, 1922, op cit; The Wall, Porth Mellin, 1923, Herb. D.-S.; dry top of hedgebank near beach, east end of Hugh Town, 1958, R. P. Bowman; Normandy, 1938, Dallas (224). Tresco: Tresco Farm garden, 1937, Herb. D.-S. St Martin's: 1958: Howitt; abundant on old stone wall above bulbfield on high ground south of Bull's Porth, 1963 (857), and 1967. Although I have never seen it in cultivation in Scilly, this handsome grass may have been grown for commercial purposes. It appears to be permanently established on wall tops at Porth Mellin, St Mary's and on St Martin's.

Sesleria Scop.

S. *caerulea* (L.) Ard. This northern grass was recorded by Townsend, 1864 in error for *Molinia caerulea*. He realised his mistake and altered his own copy of his paper.

Bromus L.

B. *sterilis* L.   Barren Brome.   Native.
Roadsides, bulbfields and waste places.   Common.
M.A.T.B.MN. : -.-.-.-.-.   First record: Townsend, 1864.

B. *madritensis* L.   Compact Brome.   Colonist.
Sandy coast and bulbfields.   Very local.
-.-.-.B.-. : -.-.-.-.   First record: Lousley, 1939.
Bryher: Southward Bay (201, var. *ciliatus* Guss. det Hubbard); bulbfield near The Town (347, var. *pubescens* Guss. det. Hubbard). This Mediterranean grass is restricted to a small area on Bryher and is probably a recent introduction.

B. diandrus Roth.   Great Brome.   Colonist.
Weed in bulbfields, roadsides.   Abundant.
M.A.T.B.MN. : -.-.-.-.-.   First record: Downes, *Rep. B.E.C.*, **7**,
413, 1924.
This is abundant in bulbfields in all the inhabited islands. It was
abundant when Downes first found it in June 1923, and although
it was not noted from St Agnes until 1952 and from Tresco until
1957, this does not necessarily reflect more recent introductions
to these islands. This Mediterranean grass has certainly been in
Scilly for 50 years and perhaps much longer.

B. mollis L.   Soft Brome.   Native.
Meadows, bulbfields, roadsides and sandy wasteland.   Common.
M.A.T.B.MN. : -.-.-.-.-.   First record: Townsend, 1864.
B. mollis in Scilly is very variable, and the small starved plants of
dry sandy places look very different from the tall luxuriant specimens in bulbfields. One very striking and elegant variation occurs
locally in bulbfields between Hugh Town and Old Town (203).
On a second gathering of this in 1940 C. E. Hubbard suggested
that the narrower flowering glumes approached *B. molliformus*
Lloyd and that it might be var. *glabrescens* Freyn of this species
of which he had seen no material. Downes recorded *B. hordeaceus*
var. *leptostachys* (Pers.) Beck from cultivated ground, St Mary's
(Thurston and Vigurs, 1922).

B. thominii Hardouin.   Lesser Soft Brome.   Native.
Dry sandy ground.   Apparently rare.
M.-.-.-.-. : -.-.-.-.E.   First record: Lousley, 1967.
St Mary's: Porthcressa beach, 1960. Eastern Isles: sandy slope,
Middle Arthur, 1957 (819 conf. A. Melderis).

B. lepidus Holmberg.   Slender Brome.   Casual.
First record: Lousley, 1967.   Has been found once—near
Rosehill, St Mary's, May 1954, R. C. L. Howitt, det. C. E. Hubbard.
Probably introduced with seed for a temporary ley.

B. racemosus L.   Smooth Brome.   Casual.
First record: Lousley, 1967.   Has been found once—in a
bulbfield between Hugh Town and Old Town, St Mary's, May
1954, R. C. L. Howitt, det. C. E. Hubbard.

B. willdenowii Kunth (*B. unioloides* auct.)   Rescue Grass.
Established alien.
Bulbfield weed, roadsides.   Locally abundant.
M.-.-.-.MN. : -.-.-.-.-.   First record: Downes in Thurston and
Vigurs, 1922.
St Mary's: roadside, Old Town, Downes, 1922; cultivated ground,
1923, Downes, *Rep. B.E.C.*, **7**, 413, 1924; bulbfield towards Old
Town, 1938 (204, teste Hubbard; f. *aristata* det. Jansen); Salakee,
1953, Polunin; Porthloo, 1956. St Martin's: bulbfield near Middle
Town, 1938 (208, teste Hubbard; f. *aristata* det. Jansen); above
Upper Town, 1963.
It seems that Rescue Grass was introduced near Old Town prior
to 1922 and spread to other parts of St Mary's, and to St Martin's.
Since 1938 it has shown a considerable though not rapid spread.
This no doubt has been helped by its very long flowering season
which extends almost through the year. I have found it in fruit as
early as the beginning of April.

Brachypodium Beauv.
B. sylvaticum (Huds.) Beauv.   Slender False Brome.   Native.
Rough coastal slopes, etc.   Common.
M.A.T.B.MN. : S.-.H.TN.E.   First record: Townsend, 1864.

Agropyron Gaertn.
A. repens (L.) Beauv.   Common Couch-grass.   Native.
Roadsides, shores and round cultivated fields.   Common.
M.A.T.B.MN. : S.-.-.-.-.   First record: Townsend, 1864.
The record for Samson is based on its occurrence on Puffin Island.
An exceedingly variable grass. Var. *barbatum* Duval-Jouve was
recorded by Somerville, 1893, and Davey, 1909 gives var. *leersianum*
S. F. Gray on the authority of Somerville. Dr C. E. Hubbard
determined var. *aristatum* Baumg., which has long-awned lemmas,
from my material from Southward, Bryher (342).

A. pungens (Pers.) Roem. and Schult. Error. Davey, 1909 quotes Townsend
as the authority for this species in Scilly, and perhaps based this on his
record of 'var. *littoreum*—Near Old Town (Townsend, 1864). In Herb.
Townsend there is a specimen from Old Town which was first labelled
'*T. pungens?* and then altered to '*repens* var.', which it is.

A. junceiforme (A. and D. Löve) A. and D. Löve.   Sand Couch.
Native.
Sandy shores, just above normal high tides.   Very common.
M.A.T.B.MN.   :   S.AT.H.TN.E.   First   record: Townsend,
1864.

Hordeum L.
  H. secalinum Schreb.   Meadow Barley.   Native.
  Marshy pastures.   Locally plentiful.   M.-.-.-.-.   :   -.-.-.-.-.
  First record: Ralfs in litt to Townsend, 1876.
  St Mary's: Old Town Marsh, Ralfs, loc cit and *Flora;* still plentiful
  there.

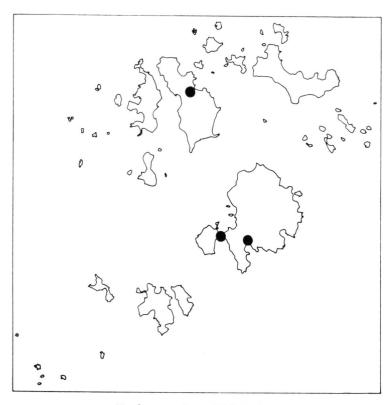

Hordeum murinum, Wall Barley

H. murinum L.   Wall Barley.   Native?
Waste places, roadsides, and especially under walls, and sandy
shores near houses.   Local.   (Map 15, p 298.)
M.-.T.-.-. : -.-.-.-.-.   First record: Ralfs, 1879.
St Mary's: Old Town, Ralfs' *Flora;* still plentiful about Old Town;
Porthcressa. Tresco: Ralfs' *Flora;* Townsend in Davey, 1909
(probably based on the record for the next species); Old Grimsby,
very local.
Although *H. murinum* is generally regarded as thriving near
the sea, the distribution does not support its status as a native.
It seems to be restricted to the vicinity of houses at Old Town,
Hugh Town and Old Grimsby which have had important shipping
trade in the past.
H. marinum Huds. This is recorded in Townsend, 1864 from 'waste
ground at New Grimsby, Tresco'. No other botanist has found it and
the species is unknown as a native in extreme south-west England. The
record must be treated as an error.

Koeleria Pers.
K. cristata (L.) Pers.   Crested Hair-grass.   Native.
Consolidated dune.   Very rare.   M.-.-.-.-. : -.-.-.-.-.
First record: Downes in Thurston and Vigurs, 1922.
St Mary's: Downes, 1922 op cit; 1939, Mrs E. Rothwell (766,
K. albescens DC. det. A. Melderis, 1957). I have failed to refind this.

K. phleoides (Vill.) Pers.   Casual.
Tresco: bulbfield weed, Northward, 1957 (771). A large form with
panicles up to 20 cm long, and interrupted. A Mediterranean
grass which might well become established.

Trisetum Pers.
T. flavescens (L.) Beauv.   Yellow Oat.   Native.
Very rare.
-.-.-.-.-. : -.-.-.TN.-.   First record: Lousley, 1967.
Tean: About a dozen plants in the grassy clearing about 30 yd north
of the pump, 1952, Grose.
The apparent restriction of this species to a small colony on one
of the uninhabited islands is extremely puzzling. Tean has been
used in fairly recent years for grazing (hence the pump) but it is a
most unlikely place for a grass to be introduced. Another puzzling
feature is that L. A. Hayward states that Yellow Oat-grass was

used in the straw-plait industry, 1800–1880, which seems unlikely unless the grass grew in greater quantity somewhere in Scilly.

Avena L.

A. fatua L.   Common Wild Oat.   Colonist.
Weed in cultivated fields.   Rare, but probably overlooked.
M.-.-.-.MN. : -.-.-.-.-.   First record: Lousley, 1967.
St Mary's: arable field near Holy Vale, 1966, M. McC. Webster (10,485), Herb. Lousley. St Martin's: Lower Town, 1962, R. C. L. Howitt (828, var. *glabrata* Peterm. det. A. Melderis).
These plants are probably descendents of oats anciently grown or introduced with the seed of other cereals.

A. nuda L. Naked Oat, Pillas, Pilcorn, Pillis, etc. *Avena nuda* is a rather primitive oat which is said to have been in cultivation in England in the thirteenth century, was extensively grown in Cornwall, and was one of the principal crops in Scilly well into the nineteenth century (Woodley, 1822, p 78). Pill (=peel, the outer coat of any sort of fruit) referred to the husk (glumes + lemma) which was deciduous at maturity. There are frequent references to Pillas by early writers on Scilly and it probably sometimes grew out of cultivation like modern oats but is apparently now extinct.

Arrhenatherum Beauv.

A. elatius (L.) Beuv. ex J. and C. Presl.   False Oat.   Native.
Roadsides and shingly beaches.   Rather rare.
M.A.-.-.-. : S.-.H.TN.E.   First record: Townsend, 1864.
St Mary's: Near Old Town, Townsend; noted recently at Tremelethen (220), London, Old Town and Porthellick. St Agnes: 1960, E. G. Philp. Samson: on Puffin Island only, Grose. St Helens. Tean: 1952, Grose. Eastern Isles: Little Arthur.
The common form in Scilly has strongly swollen tuberous stem-bases (*A. tuberosum* (Gilib.) Schultz), and plants with the stem-base not swollen have been checked by me only from a bulbfield by Old Town Lane, St Mary's. Apart from St Mary's, it is found mainly on small uninhabited islands where the plants are always tuberous.

Holcus L.

H. lanatus L.   Yorkshire Fog.   Native.
Meadows, roadsides, cliff slopes and rocky places.   Abundant.
M.A.T.B.MN. : S.AT.H.TN.E.   First record: Townsend, 1864.

Deschampsia Beauv.

D. cespitosa (L.) Beauv. Was listed by Somerville, 1893 without locality. A species quite likely to occur but the record cannot be accepted in the absence of confirmation.

Aira L.

A. praecox L.   Early Hair-grass.   Native.
Dry sandy places, rocky slopes, tops of stone walls, field borders. Very common.
M.A.T.B.MN. : S.-.H.TN.E.   First record: Townsend, 1864.

A. caryophyllea L.   Silver Hair-grass.   Native.
Dry sandy places, weed in bulbfields, wall-tops, roadsides.   Very common.
M.A.T.B.MN. : S.-.-.TN.-.   First record: Townsend, 1864.
A form with spreading prostrate stems occurs on Porthellick Beach, St Mary's (802), and on sand at Pelistry Bay, St Mary's—the latter was constant from seed in cultivation (633). I have collected similar plants from consolidated dune on the Quennevais, Jersey.

A. multiculmis Dumort.   Possibly native.
Weed in bulbfields and gardens, and on shingle.   Frequent.
M.-.T.B.MN. : -.-.-.-.-.   First record: Lousley, 1967.
St Mary's: Porthellick (211); bulbfield, Old Town (807). Tresco: Abbey Grounds (209); Borough (781). Bryher: bulbfield near Samson Hill (210). St Martin's: sandy bulbfield, Higher Town Bay (834).
This species of south-west Europe and north-west Africa seems more easily separated from *A. caryophyllea* in Scilly than elsewhere in England. There is no evident ecological separation as the two species occur close together but remain distinct. It may be an alien like some other southern grasses in bulbfields, but it is not restricted to cultivated land and it seems uniformly distributed.

Ammophila
A. arenaria (L.) Link.   Marram Grass.   Native.
Sandy coasts.   Frequent.
M.A.T.B.MN. : S.-.H.-.E.   First record: Townsend, 1864.

Marram Grass is no doubt native, but it has been so extensively planted to bind the coastal dunes that the true distribution is obscured. The inhabitants of Scilly owe a great deal to this plant. Formerly the sand from Appletree Banks and other dune areas was whipped up in gales and carried to cover whole districts like snow, and to penetrate the houses. Soon after the arrival of Augustus Smith in 1834 he initiated the planting of Marram, and later of *Carpobrotus edulis*, to bind the dunes with such success that blown sand is now only a minor nuisance.

Calamagrostis Adans.
   C. epigejos (L.) Roth.   Smallreed.   Native.

Calamagrostis epigejos, Smallreed

Rough ground on small islands.   Very local.   (Map 16, p 302.)
-.-.-.-. : -.-.H.-.E.   First record: White, 1914.
St Helen's: near Pest House, 1957. Eastern Isles: Great Ganinick,
J. W. White, *J. Bot.*, **52**, 19, 1914—still abundant at north end;
Great Ganilly, locally abundant on north-west side (315); Little
Ganilly, abundant (312); Great and Little Arthur (318).
It is remarkable that this grass, which grows on the mainland
in damp woods and fens on heavy soils, should be so plentiful on
these small rocky islands. The soils here are mainly very shallow
and sandy over granite but at least two of the colonies of Smallreed,
those on St Helen's and Great Ganilly, mark minute patches of
Head which is a little heavier in texture. The humid atmosphere
of Scilly no doubt explains why a woodland species is able to
grow in full exposure on dry ground. The colonies in Scilly are
separated by some 65 miles from the nearest known habitat on the
mainland.

Agrostis L.
A. canina L.   Velvet Bent.   Native.
Marshy meadows.   Locally plentiful.
M.-.-.-.-. : -.-.-.-.-.   First record: Townsend, 1864.
St Mary's: Old Town Marsh, Holy Vale, etc., Townsend; Higher
and Lower Moors. These are all subsp. *canina*.

A. tenuis Sibth.   Common Bent.   Native.
Heaths.   Common on the Downs.
M.A.T.B.MN. : S.-.-.TN.-.   First record: Townsend, 1864.
Townsend also recorded *A. pumila* Lightf. This is a diseased
state infected by a fungus (*Tilletia decipiens* (Pers.) Körn) and there
are specimens from Scilly in his herbarium.

A. gigantea Roth.   Black Bent.   Native.
Shady roadsides.   Rare.
M.-.T.-.-. : -.-.-.-.-.   First record: Lousley, 1940.
St Mary's: roadside near Tremelethen (257). Tresco: roadside near
Abbey House (357). Both gatherings were determined by C. E.
Hubbard as var. *dispar* (Michx.) Philipson.
On the mainland this species, with its tough widely-spreading
rhizomes, is a serious pest in arable land but, although the light
soils of Scilly might be expected to suit it, Black Bent appears to
be rare.

A. stolonifera L.   Creeping Bent.   Native.

Sandy shores, cliffs, waste ground, roadsides, marshes.   Common.
M.A.T.B.MN. : S.AT.H.TN.E.   First record: Townsend, 1864.

var. *stolonifera* is very common, and especially so on the un-inhabited islands. All the specimens in Herb. Townsend are this, including his *A. alba* var. *subrepens* and var. *coarctata*. Here also belongs *A. alba* var. *maritima* Meyer of Somerville, 1893.

var. *palustris* (Huds.) Farw. is rare and best represented by plants growing amongst *Phragmites* by Porthellick Pool, St Mary's (751, det. A. Melderis).

Apera Adans.

A. spica-venti (L.) Beauv. Loose Silky-bent. Error or casual. This is recorded as 'Scilly, Somerville' by Arthur Bennett (*J. Bot.*, **43**, Suppl. 102, 1905), but it is not mentioned in Somerville, 1893 and is likely to be an error of transcription.

Gastridium Beauv.

G. ventricosum (Gouan) Schinz and Thell.   Nitgrass.   Native.
On a roadside.   Very rare.   M.-.-.-.-. : -.-.-.-.-.
First record: Downes, in Thurston & Vigurs, 1922.

St Mary's: Old Town, 1922, Downes, loc cit; a pencil note in Downes' own copy of Davey, 1909 adds that this was found on a roadside. It is accepted as native in Cornwall and Devon, and since it is erratic in appearance I have accepted it as native here.

Lagurus L.

L. ovatus L.   Harestail Grass.   Casual.
Recorded as a garden weed at Tresco by F. T. Richards in Davey, 1903 and Davey, 1909. This is a Mediterranean species accepted as native in Guernsey but introduced in Jersey. The climate and soils of Scilly might be expected to suit it, but there is no evidence that it spread outside the garden.

Phleum L.

P. pratense L.   Timothy Grass.   Casual.
Temporary leys.   Rare.
M.-.-.-.-. : -.-.-.-.-.   First record: Lousley, 1967.

St Mary's: grass fields off Pungies Lane which had recently been reseeded, 1957, O. Moyse; temporary ley in centre of island, 1954, O. Moyse (648); temporary ley by Tremelethen, 1956 (752); field opposite Borough, 1956.

Timothy Grass is absent from Scilly as a native, but in recent years it has been introduced rather frequently in the mixtures sown for temporary leys to rest ground used for bulb cultivation.

Alopecurus L.
A. pratensis L.   Meadow Foxtail.   Casual.
Recorded by Somerville, 1893, and found on St Mary's by R. C. L. Howitt in 1954. In the absence of further details it seems likely that this is occasionally introduced in seed mixtures but is not native.

A. geniculatus L.   Marsh Foxtail.   Native.
Marshy fields and poolsides.   Local.
M.-.T.-.-. : -.-.-.-.-.   First record: Townsend, 1864.
St Mary's: Frequent in St Mary's marshes, etc., Townsend; Old Town Marsh; margin of pool near Rose Hill; pool near Watermill Cove. Tresco: School Green, 1968, P. Z. MacKenzie (spec.).
In Old Town Marsh this occurs in two forms—one with shorter, thicker spikes (394) is the one I am accustomed to seeing on the mainland; the other (395) has much slenderer spikes about 7 cm long and only about 4 mm wide.

Anthoxanthum. L.
A. odoratum L.   Sweet Vernal Grass.   Native.
Meadows, roadsides, edges of bulbfields, and heathy ground on coastal slopes.   Very common.
M.A.T.B.MN. : S.-.H.TN.E.   First record: Townsend, 1864.

Phalaris L.
P. canariensis L.   Canary Grass.   Alien.
Cultivated ground.   Rare.
M.-.T.-.-. : -.-.-.-.-.   First record: Boyden, 1890.
St Mary's: Hugh Town, Smith, 1909; field near Parting Carn, 1928, Herb. D.-S. Tresco: Smith, 1909. Also recorded (without locality) by Tellam in Davey, 1909, and by Downes from 'cultivated ground' in *Rep. B.E.C.*, **7**, 410, 1924.
This species has been much confused with *P. minor* to which some of the above records probably belong. Downes' material is correctly named (Herb. S.L.B.I.) and it is likely that this species is being currently introduced with bird-seed mixtures.

P. minor Retz.   Lesser Canary Grass.   Naturalised alien.
Weed in bulbfields.   Locally abundant.
M.-.-.-.MN. : -.-.-.-.   First record: Lousley, 1939.
St Mary's: bulbfield near Rocky Hill (221, teste C. E. Hubbard);
bulbfield, Church Road; field entrance near Parting Carn; side of
road to Peninnis Head, 1953, Polunin; bulbfield, Old Town Lane
(656; 804). St Martin's: 1962, Howitt.
This species has spread very slowly in the last 30 years. It is a
Mediterranean grass which occurs up the west coast of France to
Normandy and is accepted as native in Guernsey.

Nardus L.
N. stricta L. Error? Recorded from St Mary's in Ralfs' *Flora*, 1879.
This might well occur but records from Ralfs cannot be accepted in the
absence of confirmation.

Cynodon Rich.
C. dactylon (L.) Pers.   Dogstooth Grass.   Bermuda Grass.
Naturalised alien.
In short turf on sandy ground near sea.   Locally abundant.
-.-.T.-.-. : -.-.-.-.   First record: Lousley, 1967.
Tresco: abundant on School Green, 1956 (716), 1970. This grass
has been known at Marazion since 1688 and the original habitat
there was very similar to the one in Scilly. It also has the appearance
of a native plant in the Channel Isles. Except when it grows on
bare ground, Bermuda Grass is easily overlooked and it must have
been on School Green for many years before its discovery.

Digitaria P. C. Fabr.
D. sanguinalis (L.) Scop.   Crab Grass.   Naturalised alien.
Sandy shore, and weed in gardens.   Rare.
-.-.T.-.-. : -.-.-.-.   First record: Lousley, 1939.
Tresco: On the shore near Farm, Herb. D.-S.; weed in Abbey
Grounds (353, teste C. E. Hubbard).

Setaria Beauv.
S. verticillata (L.) Beauv.   Rough Bristle-grass.   Alien.
Garden weed.   Very rare.
-.-.T.-.-. : -.-.-.-.   First record: Lousley, 1940.
Tresco: weed in Abbey Grounds (352) teste C. E. Hubbard).

Arundinaria Michx.

A. japonica Sieb. and Zucc.　　Bamboo.　　Established alien.
Local.

M.-.-.-.-. : -.-.-.-.-.　First record: this *Flora*.

St Mary's: Toll's Island, in bracken, 1969, R. Lancaster; Rocky
Hill, 1970.

BIBLIOGRAPHY

A great many general works on the British flora include occasional references to plants found in Scilly. These are not included here but are cited in the text where appropriate.

Air Ministry, Meteorological Office (1938). *Averages of humidity for the British Isles.*

Air Ministry, Meteorological Office (1952). *Climatological Atlas of the British Isles.*

Anon (1872). 'The Crown of Scilly', *Gdnrs' Chron.* for 1872, 1102–3.

Anon (1872). 'List of some of the plants growing all the year round in the ground of the late Mr Augustus Smith, at Tresco, in the Scilly islands', *Gdnrs' Chron.* for 1872, 1129.

Anon (1896). 'The Flower Trade at Scilly', *The Times* for 29 May, 1896, 8.

Anon (1970). 'Scarcer Maps and Books of Scilly', *Isles of Scilly Museum Publ.* No 8, Scilly.

Ashbee, Paul (1955). 'Excavation of a Homestead of the Roman Era at Halangy Down, St Mary's, Isles of Scilly, 1950', *Antiquaries' Jl.*, 35, 187–198.

Barrow, George (1906). *The Geology of the Isles of Scilly.*

Beeby, W. H. (1873). Letter to F. Townsend dated 27 March, 1873 at South London Botanical Institute, Norwood, London.

Bilham, E. G. (1938). *The Climate of the British Isles.*

Blair, K. G. (1925). 'The Lepidoptera of the Scilly Isles', *Entomologist* 58 (3–10 and 38).

Blair, K. G. (1931). 'The Beetles of the Scilly Isles', *Proc. zool. Soc. Lond.*, 1211–58.

Borlase, Wm. (1754). 'An Account of the great Alterations which the Islands of Sylley have undergone since the time of the Antients, who mention them, as to their Number, Extent and Position', *Phil. Trans. R. Soc.*, 48, 55–68.

Borlase, Wm. (1756). *Observations on the Ancient and Present State of the Islands of Scilly . . .*, Oxford. (Also Ed 2, 1966. Newcastle).

Bowles, E. A. (1934). *A Handbook of Narcissus.*

Bowley, E. L. (1939). *Illustrated Guide (to the) Isles of Scilly*, Isles of Scilly.

Bowley, E. L. (1945). *The Fortunate Isles*, Ed 2, 1947; Ed 3, 1949; Ed 4, 1957; Ed 5, rewritten by R. L. Bowley, 1964, Reading.

Bowley, R. L. (1970). *The Standard Guidebook to the Isle of Tresco*, Reading.

Boyden H. (1890). 'The Flora of the Scilly Islands' (summary of paper read January 10, 1890), *Trans. Penzance Nat. Hist. and Antiq. Soc.*, 3 (N.S.), 186.

Boyden, H. (1893). 'The Flora of the Isles of Scilly', in Tonkin and Row, 1893, 122–6.

Boyden, H. (1906). 'The Flora of the Isles of Scilly', in Tonkin and Row, 1906, 126–9.

Bree, W. T. (1821)—see Forbes, 1821.

Bree, W. T. (1831). 'List of rare plants found in the Neighbourhood of Penzance' (includes 12 records from Scilly, of which 10 are repeated in Watson, 1835), *Loudon's Mag. Nat. Hist.*, 4, 161–2.

Brewer, E. (1890). 'Market-gardening in the Scilly Islands', *J. Roy. Agric. Soc. (England)*, 1 (ser. 3), 219–23.

Bristowe, W. S. (1929). 'The Spiders of the Scilly Islands', *Proc. zool. Soc. Lond.*, 1929, Part 2, 149–64.

Bristowe, W. S. (1929b). 'A contribution to the knowledge of the Spiders of the Channel Islands', *Proc. zool. Soc. Lond.*, 1929, 181–8.

Bristowe, W. S. (1929c). 'The Spiders of Lundy Island', *Proc. zool. Soc. Lond.*, 1929, 235–44.

Bristowe, W. S. (1935). 'Further Notes on the Spiders of the Scilly Islands', *Proc. zool. Soc. Lond.*, 1935, 219–32.

Bristowe, W. S. (1958). *The World of Spiders.*

Brown, E. (1935). 'Tresco Abbey', *Gdnrs' Chron.*, 98 (ser. 3), 102–3.

Carmichael, C. A. M. (1884). The Gardens at Tresco Abbey. *The Garden*, 26, 333–5.

Clapham, A. R., Tutin, T. G., Warburg, E. F. (1962). *Flora of the British Isles*, Ed 2, Cambridge.

Clapham, A. R., Tutin, T. G., Warburg, E. F. (1968). *Excursion Flora of the British Isles*, Ed 2, Cambridge.

Cooke, E. W. (1850). 'Ferns'. in North 1850, 183–6.

Cooke, E. W. (1867). 'Ferns', in Courtney, 1867.

Cooke, E. W. (1882). 'Ferns', in Courtney, 1882, 68–9.

Cosmo III, Duke of Tuscany (c1669)—see Magalotti, Count.

Courtney, J. S. (1845). *A Guide to Penzance . . . including the Islands of Scilly.*

310   BIBLIOGRAPHY

Courtney, Leonard H. (1867). *A week in the Isles of Scilly*, Truro.

Courtney, Leonard H. (1883). *A week in the Isles of Scilly*. This is a second edition of the 1867 work greatly revised but the cover is printed 'Guide to the Scilly Islands' with the date 1882. Truro.

Curnow, W. (1876). 'A Botanical Trip to the Scilly Isles', *Hardwicke's Science Gossip*, 1876, 162.

Curnow, W. (1882). 'List of Flowering Plants', in Tonkin, 1882.

Curnow, W. (1893). 'List of Flowering Plants', in Tonkin, 1893.

'Daffodil Handbook' (1966) in *Amer. Hort. Mag.*, 45, 1–227.

Dallas, J. E. S. (1938). Manuscript notes of a visit in June 1938 in author's possession.

Dallimore, W. (1913). 'The culture of early flowers in Cornwall and the Scilly Islands', *Kew Bull.*, 1913, 171–7.

Dandy, J. E. (1958). *List of British Vascular Plants.*

Daubeny, C. (1863). *Climate, an Inquiry into the causes of its differences and into its influence on vegetable life.*

Davey, F. Hamilton (c1901). 'Contributions to a Cornish Flora', *J. Ryl. Instit. Cornwall* 48.

Davey, F. Hamilton (1902). *A Tentative List of the Flowering Plants, Ferns, etc of Cornwall*, Penryn.

Davey, F. Hamilton (1903). 'Further Contributions to the Cornish Flora', *J. Ryl. Instit. Cornwall*, 50.

Davey, F. Hamilton (1904). 'Botanical Report for 1904', *J. Ryl. Instit. Cornwall.* 51.

Davey, F. Hamilton (1905). 'A Botanical Report for 1905', *J. Ryl. Instit. Cornwall*, 52.

Davey, F. Hamilton (1906). 'A Botanical Report for 1906', *J. Ryl. Instit. Cornwall*, 53.

Davey, F. Hamilton (1907). 'Additions to the Flora of Cornwall', *J. Ryl. Instit. Cornwall*, 54.

Davey, F. Hamilton (1910). *Flora of Cornwall . . . including the Scilly Isles*, Penryn.

Davey, F. Hamilton (1910). 'A Botanical Report for 1909, 1910', *J. Ryl. Instit. Cornwall*, 57.

Davey, F. Hamilton (1906). Botany, in *Vict. Cty. Hist. Cornwall*, 1, 49.

Dimbleby, G. W. (1967). *Plants and Archaeology.*

Dimbleby, G. W. (1967a). *Vegetation History of the Isles of Scilly, an interim report on pollen analyses.* Manuscript on loan to author.

Dorrien-Smith, A. A. (1908). The Southern Islands Expedition. *Kew Bull.*, 1908, 237–49.

Dorrien-Smith, A. A. (1910). 'A botanising expedition to West Australia in the Spring (October) 1909', *Jl. R. hort. Soc.*, 36, 285–93.

Dorrien-Smith, A. A. (1911). 'Plants of Chatham Island', *Jl. R. hort. Soc.*, 37, 57–64.

Dorrien-Smith, A. A. (1926). 'Flower growing for market in the Isles of Scilly since the Great War', *Jl. R. hort. Soc.*, 51, 266–70.

Dorrien-Smith, A. A. (undated). *Tresco Abbey Gardens*, Tresco.

Dorrien-Smith, Gwen (1953). 'List of Wildflowers', in Vyvyan, 1953, 154–60.

Dorrien-Smith, T. A. (1890). The progress of the Narcissus Culture in the Isles of Scilly', *Jl. R. hort. Soc.*, 12, 311–16.

Downes, A. (1957). 'Farming the Fortunate Islands', *Geography*, 42, 105–12.

Downes, H. (1920, 1921). Marginal notes in a copy of Davey, 1910 given to the author by his widow.

D'Oyly, M. J. (1959). *Some Plant Communities on Tresco and Samson.* Typescript. Copy in author's possession.

Elkan, E. (1954). 'The Succulent West', *Nat. Cactus and Succ. J.*, 9, 49–53.

Fernandes, A. (1968). 'Key to the identification of Native and Naturalised Taxa of the Genus *Narcissus* L.', *Daffodil and Tulip Year Book* (R. hort. Soc.), 33, 37–66.

Fitzherbert, S. W. (1898). 'An August Visit to Tresco Abbey Gardens. *The Garden*', 54, 473–4.

Fitzherbert, S. W. (1902). 'Early March in Tresco Abbey Gardens'. *The Garden*, 61, 227–8.

Forbes, John (1821). *Observations on the Climate of Penzance and the District of the Lands' End in Cornwall, with an Appendix containing Meteorological Tables, and a catalogue of the rarer indigenous plants* [by W. T. Bree], Penzance.

Ford, E. B. (1967). *Butterflies*, Ed 3.

Forfar, Rebecca (1896). *Evenings with Grandmama—Recollections of the Scilly Islands*, Truro. (The British Museum copy is inscribed 'In memory of the writer. Elizth. F. Ellis').

Gerrans, M. B. (1967). 'Flora of the Scillies', *Gdrns' Chron.*, 161, (10). 22–6.

Gibson, A. G. and H. J. (1932). *The Isles of Scilly*, Ed 2, Scilly.

Gibson, Gordon W. (1934). 'The Early Days of the Scillonian Flower Industry', *Daffodil Year-book* (R. hort. Soc.), 1934, 24–9.

Gibson and Sons, (*c* 1931). *Guide to the Scilly Isles*, Scilly.

Gillham, M. E. (1953). 'An ecological account of the vegetation of Grassholm Island, Pembrokeshire', *J. Ecol.*, 41, 84–99.

Gillham, M. E. (1955). 'Ecology of the Pembrokeshire Islands 3. The effect of grazing upon the vegetation', *J. Ecol.*, 43, 172–206.

Gillham, M. E. (1956a). 'Ecology of the Pembrokeshire Islands 4. The effects of treading and burrowing by birds and mammals', *J. Ecol.*, 44, 51–82.

Gillham, M. E. (1956b). 'Ecology of the Pembrokeshire Islands 5. Manuring by the colonial seabirds and mammals, with a note on seed distribution by gulls', *J. Ecol.*, 44, 429–54.

Girard, P. J. (1967). *A History of the Bulb and Flower Industry in Guernsey*, States Committee for Horticulture, Guernsey.

Goldsmith, F. B., Munton, R. J. C., and Warren, A. (1970). 'The impact of recreation on the ecology and amenity of semi-natural areas: methods of investigation used in the Isles of Scilly', *Biol. J. Linn. Soc.*, 2, 287–306.

Goodman, G. T. and Gillham, M. E. (1954). 'Ecology of the Pembrokeshire Islands 2. Skokholm, environment and vegetation', *J. Ecol.*, 42, 296–327.

Graves, P. P. (1930). 'The British and Irish Maniola jurtina L. (L.E.P.)', *Entomologist*, 63, 75–81.

Gray, Alex (1933). 'Where our Flowers come from', *Scillonian* 7, 22–4.

Green, M. A. E. (Edit) (1871). *Calendar of State Papers, Domestic Series, of the reign of Elizabeth. Addenda* 1566–1579, 559.

Grigson, G. (1948). *The Scilly Isles*. (An outstanding scholarly work).

Grigson, G. (1956). 'A little-known Isle of Scilly (White Island, St Martin's)', *Country Life*, 119, 374–5.

Grose, J. D. (1939) and (1952). Typed copies of detailed observations made on visits in July 1939 and July 1952. In possession of author.

Guppy, H. B. (1925). 'A Side-issue of the Age-and-Area Hypothesis', *Ann. Bot.*, 39, 805–8.

Hadfield, Miles (1957). 'Strange Trees of Seaside Gardens', *Gdnrs' Chron.*, 141, 3.

Haley, W. B. (1921). 'Plants of the Scilly Isles', *Naturalist*, 1921, 328.

Heath, Robert (1750). *A Natural and Historical Account of the Islands of Scilly*. (Also cheap reprint, 1967.)

Hencken, H. O'Neil (1932). *The Archaeology of Cornwall and Scilly*.

Hill, A. W. (1920). 'Tresco Abbey Gardens, Scilly Isles', *Kew Bull.*, 1920, 170–4.

Holden, M. (1965). 'Spring Foray, St Mary's, Isles of Scilly', *Br. mycol. Soc. News Bull.*, 24.

Hooker, J. D. (1872). '*Olearia dentata*', *Curtis's bot. Mag.*, tab 5973.

Hooker, J. D. (1902). 'A Sketch of the Life and Letters of Sir William Jackson Hooker', *Ann. Bot.*, 16, ix–ccxxi.

Hooker, W. J. (1813). Letters to Dawson Turner dated 26 March, 1813 and 27 April, 1813 in library of *Roy. Bot. Gardens, Kew*.

Hunkin, J. W. (1947). 'Tresco under three reigns', *Jl. R. hort. Soc.*, 72, 177–191, 221–37.

Hyde, H. A. (1956). 'Tree Pollen in Great Britain', [Catches at Bishop Rock]. *Acta allergologica*, 10, 224–45.

Hyde, H. A. (1959). ' "Weed" Pollen in Great Britain' [catches at Bishop Rock], *Acta Allergologica*, 13, 186–209.

Hyde, H. A. and Adams, K. F. (1961). *Spore Trapping in relation to Epidemics of Black Rust* 1947–59 [catches at Bishop Rock]. Duplicated typescript prepared for Il Coloquio Europeo sobre la Roya Negra de los Cereales, Madrid, 1961.

Inglis-Jones, E. (1969). *Augustus Smith of Scilly*.

Jacob, J. (1835–7). *Flora of West Devon and Cornwall*.

Jefferson-Brown, M. J. (1969). *Daffodils and Narcissi*.

Kay, E. (1956). *Isles of Flowers*.

King, C. J. (1924). *Some Notes on Wild Nature in Scillonia*, Scilly.

Knox, M. (1938). Manuscript notes communicated to the author.

Lawson, M. A. (1870). 'Additions to the Flora of the Scilly Islands', *J. Bot.*, 8, 357–8.

Leland, John (1710). *The Itinerary of John Leland* 1535–1543, First published 1710. Quoted from edition edited by L. T. Smith, 1907, London.

Leland, John (1964). *The Itinerary of John Leland in or about the years* 1535–1543. Ed Lucy Toulmin Smith, Vol 1.

Leuze, Eva (1966). 'Die Scilly Inseln', *Erdkunde*, 20, 93–103.

Lewes, G. H. (1859). *Scilly Inseln und Jersey: Naturstudien am Seestrande. Küstenbilder aus Devonshire*. Trans. J. Frese. (Not seen).

Lewes, George Henry. (1860). *Sea-side Studies at Ilfracombe, Tenby, The Scilly Isles and Jersey*, Ed 2, Edinburgh and London.

Lewis, Agnes G. (1907). *John Ralfs. An old Cornish botanist*, Torquay.

Lousley, J. E. (1939). 'Notes on British Rumices 1', *Rep. Bot. Soc. and E.C.*, 12, 118–57; (1944) id 2. *Rep. Bot. Soc. and E.C.*, 12, 547–85.

Lousley, J. E. (1939b). 'Notes on the Flora of the Isles of Scilly—1', *J. Bot.*, 77, 195–203.; (1940) id.—2, *J. Bot.*, 78, 153–60.

Lousley, J. E. (1944). 'A Botanical Hotel (Flora of Star Castle)', *Rep. Bot. Soc. and E.C.*, 12, 527.

Lousley, J. E. (1967). *Flowering Plants and Ferns in the Isles of Scilly*, (1970) Ed 2; Isles of Scilly Museum Publ. 4, Scilly.

Magalotti, Lorenzo (1821). *Travels of Cosmo the Third, Grand Duke of Tuscany through England during the reign of King Charles the Second, 1669.*

Manley, Gordon (1952). *Climate and the British Scene.*

Marquand, E. D. (1890). 'The Flora of Guernsey compared with that of West Cornwall', *Trans. Penzance Nat. Hist. and Antiq. Soc.*, 3 (N.S.), 132–40.

Marquand, E. D. (1893). 'Further Records for the Scilly Isles', *J. Bot.*, 31, 265–7. [The records in this paper are taken from Ralfs' *Flora*, 1879, and I have preferred to quote from the original manuscript.]

Matthews, G. Forrester (1960). *The Isles of Scilly: a Constitutional, Economic and Social Survey of the Development of an Island People from early times to 1900.*

Matthews, J. R. (1937). 'Geographical Relationships of the British Flora', *J. Ecol.*, 25, 1–90.

Matthews, J. R. (1955). *Origin and Distribution of the British Flora.*

Meteorological Office (1963). *Averages of Temperature for Great Britain and Northern Ireland* 1931–1960.

Millett, L. and M. (1853). 'Wild Flowers and Ferns of the Isles of Scilly observed in June and July', *Trans. Nat. Hist. and Antiq. Soc. Penzance*, 2, 75–8.

Min. Agric. Fish and Food (1967). *Agricultural Statistics* 1964/65. *England and Wales.*

Mitchell, G. F. (1960). 'The Pleistocene History of the Irish Sea', *Advmt. Sci., Lond.*, 17, 313–25.

Mitchell, G. F. and Orme, A. R. (1967). 'The Pleistocene deposits of the Isles of Scilly', *Q. Jl. geol. Soc. Lond.*, 123, 59–92.

Mitchinson, J. G. (1881). 'Tresco Abbey', *Gdnrs' Chron.*, 1881, 84.

Montgomery, J. B. (1854). 'List of Phaenogamous Plants and Ferns of Western Penrith', *Trans. Nat. Hist. and Antiq. Soc. Penzance*, 2, 215–22.

Moore, W. Robert (1938). 'The Garden Isles of Scilly', *Nat. Geogr. Mag.*, 74, 755–74.

Mortimer, Maryellen (1941). 'Notes on Scilly and the Factors influencing land use in the Islands', in Robertson, 1941.

Mothersole, Jessie (1919). *The Isles of Scilly*, Ed 3, (Ed 1, 1910; Ed 2, 1914).

Mumford, Clive (1967). *Portrait of the Isles of Scilly*.

(Murray, John) (1872). *Handbook for Travellers in Devon and Cornwall*. Ed 8, 470.

North, I. W. (1850). *A Week in the Isles of Scilly*, Penzance and London.

O Neil, B. H. St. J. (1952). 'The Excavation of Knackyboy Cairn, St Martin's, Isles of Scilly, 1948', *Antiquaries' Jl.*, 32, 21–34.

O'Neil, B. H. St. J. (1954). 'A triangular cist in the Isles of Scilly', *Antiquaries' Jl.*, 34, 235–7.

O'Neil, B. H. St. J. (1969). *Ancient Monuments of the Isles of Scilly*, Ed 2 (of 1961), Fourth impression with amendments, (Ed 1, 1949).

Osman, C. W. (1928). 'The Granites of the Scilly Isles and their relation to the Dartmoor Granites', *Q. Jl. geol. Soc. London.*, 84, 258–92.

Owen, James G. (1897). *Faire Lyonesse: A Guide to the Isles of Scilly*, Bideford.

'B.P.' (1884). Ten Days in Scilly, *Science Gossip*, 20, 98.

Paton, Jean A. (1968). Wild Flowers in Cornwall, Truro.

Perring, F. H. (1968). *Critical Supplement to the Atlas of the British Flora*.

Perring, F. H. and Walters, S. M. (1962). *Atlas of the British Flora*.

Plumtre, C. J. (1890). Notebook. MS. with records of a visit in April 1890. Lent by D. McClintock.

Polunin, O. (1953). *Some Plant Communities of the Scilly Isles*. Duplicated typescript—copy in author's possession.

Praeger, R. Ll. (1911). Clare Island Survey, Part 10. Phanerogamia and Pteridophyta', *Proc. R. Ir. Acad.*, 31.

Ralfs, John (1876a). Letter to F. Townsend dated 2 October, 1876 in library of South London Botanical Institute.

Ralfs, John (1876b). *Materials towards a Flora of the Scilly Islands.* Manuscript notebook now in Penzance Library, Morrab Gardens, Penzance. This seems to have been prepared as an aide-memoire to be carried in the field. As it is incorporated with many additions and corrections in his *Flora of West Cornwall*, 1879, I have preferred to quote from the later manuscript.

Ralfs, John (1877). Letter to F. Townsend dated 7 July, 1887 in library of South London Botanical Institute.

Ralfs, John (1879). *Flora of West Cornwall (West Penrith)*, vol 3, p 219–80. Manuscript in Penzance Library, Morrab Gardens. This is a compilation and it is difficult to decide which of the new records are correctly attributable to Ralfs.

Ranwell, D. S. (1966). 'The Lichen Flora of Bryher, Isles of Scilly, and its Geographical components', *Lichenologist*, 3, 224–32.

Raven, J. E. (1950). 'Notes on the flora of the Scilly Isles and the Lizard Head', *Watsonia*, 1, 356–8.

Rees, Edgar A. (1932). 'Flora of Scilly'. in Gibson, 1932 p 61–4.

Ribbons, B. W. (1953). '*Acanthus mollis* L. in St Agnes, Isles of Scilly', *Watsonia*, 2, 392–3.

Robertson, B. S. (1941). 'Isles of Scilly', in *The Land of Britain* (Report of Land Utilisation Survey of Britain), Ed L. Dudley Stamp, 91 (Cornwall), 407–67.

Salmon, E. S. (1893). 'A summer holiday in Cornwall and the Scilly Isles', *Proc. Holmesdale Nat. Hist. Soc.*, 1890–92, 35.

Sandwith, C. (1939). Manuscript notes on plants seen on several visits sent to the author.

Savage, S. (1925). *Memorandum on the Survival of Diotis candidissima Desf. in the United Kingdom.* Typescript in possession of author.

Savonius, M. (1960). 'A sub-tropical Garden in Britain (Tresco)', *Gdnrs' Chron.*, 148, 494 seq.

Scott, L. and Rivington, H. (1870). 'The Agriculture of the Scilly Isles', *Jl. R. agric. Soc.*, 6 (Ser. 2), 374–92.

Shepperd, F. W. (1960). 'Trees and Shrubs for Shelter', *Jl. R. hort. Soc.*, 85, 17–26.

Smith, Augustus (1849, 1865, 1866, 1871). Letters to W. J. and J. D. Hooker in the Library of the Roy. Bot. Gardens, Kew.

Smith, J. D. H. (1964). 'Tresco Abbey Gardens', *N. Wales Gardener*, 9, 11–17.

Smith, W. W. (1907). 'Note on a Peculiar Tussock Formation', *Trans. bot. Soc. Edinb.*, 23, 234–5.

Smith, W. W. (1909). 'Notes on the Flora of the Scilly Isles', *Trans. bot. Soc. Edinb.*, 24, 36–8. (Records of plants noted on a visit in 1906.)

Somerville, A. (1893). 'Additional Records for the Scilly Isles', *J. Bot.*, 31, 118–20. [This is based on a visit made in 1890. No localities are given, and 10 of the species recorded are still unconfirmed.]

Southern, H. N. (Ed) (1964). *The Handbook of British Mammals*, Oxford.

Spooner, D. C. (undated). *The Lost Land of Lyonesse.* Typescript.

Steers, J. A. (1948). *The Coastline of England and Wales*, Cambridge.

Stockdale, F. W. L. (1824). *Excursions in the County of Cornwall.*

Suffern, C. (1939). 'Letter citing ancient deeds in Public Record Office,' *British Birds*, 32, 376.

Taylor, Thos. (1906). 'Industries (of Cornwall)', *Vict. County Hist. Cornwall*, 1, 578–80.

Teague, A. Henwood (1891). 'Plants growing in Tresco Abbey Gardens', *Trans. Penzance Nat. Hist. and Antiq. Soc.*, 3 (N.S.), 157–71.

Thurston, E. (1929). 'Note on the Cornish Flora' (read 1928), *J. Roy. Instit. Cornwall*, 23, 74–86.

Thurston, E. (1936). 'Flowering Plants and Ferns (of Cornwall) 1930–1934', *J. Roy. Instit. Cornwall*, 24, 240–86.

Thurston, E. and Vigurs, C. C. (1922a). 'A Supplement to F. Hamilton Davey's Flora of Cornwall', *J. Roy. Instit. Cornwall*, 21, i–xx, 1–172.

Thurston, E. and Vigurs, C. C. (1922b). 'Note on the Cornish Flora', *J. Roy. Instit. Cornwall*, 21, 164–8.

Thurston, E. and Vigurs, C. C. (1926). 'Note on the Cornish Flora', (read 1925), *J. Roy. Instit. Cornwall*, 22, 99–112.

Thurston, E. and Vigurs, C. C. (1927). 'Note on the Cornish Flora' (read 1926), *J. Roy. Instit. Cornwall*, 22, 252–68.

Thurston, E. and Vigurs, C. C. (1928). 'Note on the Cornish Flora' (read 1927), *J. Roy. Instit. Cornwall*, 22, 350–63.

Todd, Miss E. S. (1933, 1934). Manuscript lists of plants observed in 1933 and 1934 communicated to the author.

Tonkin, J. C. and R. W. (1882). *Guide to the Isles of Scilly* (1887), Ed 2, Penzance.

Tonkin, J. C. and R. W. (1893). *Guide to the Isles of Scilly*, Ed 3. Penzance. (Lists of flowering plants by Curnow and Boyden.)

Tonkin, J. C. and Row, Prescott (1906). *Lyonesse: A Handbook for the Isles of Scilly*, Ed 4, Scilly and London. (List of plants by Boyden.)

[Towers, Lady Sophia] (1849). *Sketches in the Isles of Scilly*, London, Exeter and Penzance.

Townsend, F. (1862). *Catalogue of Plants—Scilly Isles—June* 1862. Manuscript from which the 1864 printed account was prepared. Library of South London Botanical Institute.

Townsend, F. (1864). Contributions to a Flora of the Scilly Isles. *J. Bot.*, 2, 102–20.

Townsend, F. (MS.). Townsend's copy of his paper in the Journal of Botany with many corrections and additions, two letters from John Ralfs and one from W. H. Beeby inserted. Library of South London Botanical Institute.

Tremayne, Lawrence J. Notes on Cornish plants in an interleaved copy of the *London Catalogue of British Plants*, ed 10, in possession of author.

Troutbeck, J. (1794). *A Survey of the Ancient and Present State of the Scilly Islands*, Sherborne.

Tucker, Benjamin (1810). *The Report of the Surveyor-General of the Duchy of Cornwall . . . concerning the formation of a Safe and Capacious Roadstead within the Islands of Scilly*.

Turner, Sir George (1964). 'Some Memorialls towards a Natural History of the Sylly Islands (MS c1695)', *Scillonian*, no 159, 153–7.

T. W., H. (1895). 'Ten Days in Lilyland', *Gdnrs' Chron.*, 17 (ser. 3), 389–90.

Vagabondo (1886). 'Tresco', *Gdrns' Chron.*, 26 (N.S.), 558.

Van den Brink, F. H. (1967). *A Field Guide to the Mammals of Britain and Europe*.

Vyvyan, C. C. (1954). *The Scilly Isles*.

Wace, N. M. and Dickson, J. H. (1965). 'The Terrestial Botany of the Tristan da Cunha Islands, *Phil. Trans. R. Soc. Ser. B.*, 249, 273–360.

Wanstall, P. J. and Ribbons, R. W. (1948). Typed list of species observed on St Agnes.

Ward, Rodney W. (1964). 'Early Daffodils in the Isles of Scilly', *The Daffodil and Tulip Year Book* for 1965, 30, 54–5.

Watson, H. C. (1835). *New Botanists' Guide*, Vol 1, 6–12.

Wheeler, R. E. M. (1941). 'Hill-forts of Northern France: a note on the expedition to Normandy, 1939', *Antiquaries' Jl.*, 21, 265–70.

White, J. W. (1914). Plants of Scilly. *J. Bot.*, 52, 19.

Whitfield, H. J. (1852). *Scilly and its Legends.*

Wilkinson, C. (1955). 'Spring Blooms in the Scillies', *Field*, 205, 220–1.

Williams, Thomas (1900). *Life of Sir James Nicholas Douglass.*

Willis, J. C. (1922). *Age and Area*, Cambridge.

Woodley, Geo. (1822). *A view of the Present State of the Scilly Islands.*

Woods, Jos. (1852). 'Letter with plant records', *Trans. Nat. Hist. and Antiq. Soc. Penzance*, 2, 100.

ACKNOWLEDGEMENTS

It has taken over 30 years to collect the material for this *Flora* and it would be impractical to list all those to whom I am indebted for help. I trust that those who are not mentioned individually will understand that this is not on account of any lack of gratitude.

The names of those who contributed plant records appear against the individual records but special mention must be made of J. D. Grose, the late Mr and Mrs J. E. Dallas, and the late Miss Margaret Knox. Their work at the routine task of island by island recording was so complete by 1940 that surprisingly little has been added to the basic distribution since. Special mention too must be made of R. P. Bowman, Miss Ann Conolly, Mr and Mrs R. C. L. Howitt, Roy Lancaster and J. E. Raven for especially useful contributions. Many specialists have given their help and their names appear against the records of the specimens identified.

This book could not have been written without the facilities afforded to me at many libraries and herbaria and especially those at the British Museum; British Museum (Natural History); Royal Botanic Gardens, Kew; Bodleian Library, Oxford; Guildhall Library, City of London; and South London Botanical Institute. The librarian at Truro Public Library has kindly helped with references to Cornish periodicals. To David McClintock, B. T. Ward and others I am grateful for drawing my attention to certain specimens in herbaria, and to the staff of the Distribution Maps Scheme, and Biological Records Centre of the Nature Conservancy, for advising me of records of plants from Scilly they received.

As a non-resident I have been dependent on local help over many practical aspects and I am especially grateful to the late Major Arthur Dorrien-Smith, Mrs C. Harvey, P. Z. MacKenzie, Mrs O. R. Moyse, and Miss Hilda Quick.

I am indebted to the Publications Committee of the Botanical Society of the British Isles for being patient with a dilatory author for so many years, and to Franklyn Perring whose initiative has led to the present publication. Keeping the manuscript abreast of repeated changes in nomenclature has been one of my greatest problems, and I am especially grateful to J. E. Dandy and Dr W. T. Stearn for their help.

The Nature Conservancy have kindly provided the revised acreages of the islands as set out in Table 1 and the Meteorological Office, Bracknell climatic statistics for Scilly and Jersey. Tables 2, 3 and 5 are reproduced with the permission of the Copyright Division, H.M. Stationery Office, Crown Copyright reserved. To Prof G. W. Dimbleby I am most grateful for permission to quote from an unpublished manuscript which he sent me on loan.

The pages of this *Flora* bear witness to how much I owe to the columns of *The Scillonian*, in which residents faithfully record local affairs quarter by quarter and visiting authorities provide contributions on their special subjects. To Barbara Everard I am grateful for her care and skill in preparing the drawings. some of them of plants not previously illustrated in a British Flora. To my wife Dorothy, who shares all my enthusiasm for Scilly, I owe the drawings for the other diagrams and much help and encouragement over the years.

INDEX

Figures in heavy type refer to illustrations. Figures followed by an asterisk refer to pages on which Distribution Maps appear. Scientific names are given in full for illustrations and maps but generic names only are indexed for pages 91 to 307.

324 INDEX

D1